Authority and Imitation

Mittellateinische Studien und Texte

VOLUME 47

The titles published in this series are listed at *brill.com/mits*

Authority and Imitation

A Study of the Cosmographia *of Bernard Silvestris*

By

Mark Kauntze

BRILL

LEIDEN | BOSTON

Library of Congress Cataloging-in-Publication Data

Kauntze, Mark.
Authority and imitation : a study of the Cosmographia of Bernard Silvestris / by Mark Kauntze.
pages cm. — (Mittellateinische Studien und Texte, ISSN 0076-9754 ; volume 47)
Includes bibliographical references and index.
ISBN 978-90-04-25691-0 (hardback : acid-free paper) — ISBN 978-90-04-26835-7 (e-book)
1. Bernard Silvestris, active 1136. Cosmographia. 2. Bernard Silvestris, active 1136—Criticism and interpretation. 3. Literature, Medieval—History and criticism. 4. Didactic poetry, Latin (Medieval and modern)—History and criticism. 5. Cosmology, Medieval—Poetry. 6. Tours (France)—Intellectual life. I. Title.

PA8275.B25C634 2014
189—dc23

2014008070

This publication has been typeset in the multilingual 'Brill' typeface. With over 5,100 characters covering Latin, IPA, Greek, and Cyrillic, this typeface is especially suitable for use in the humanities. For more information, please see brill.com/brill-typeface.

ISSN 0076-9754
ISBN 978 90 04 25691 0 (hardback)
ISBN 978 90 04 26835 7 (e-book)

This book is printed on acid-free paper.

For Cathy

Contents

Acknowledgements

I would like to thank the Read-Tuckwell Foundation for funding my doctoral research at the University of Bristol, and Ian Wei and James Clark for providing excellent supervision, which combined criticism and encouragement in just the right measures. I am also grateful to John Marenbon for suggesting the connections between the *Cosmographia* and the schools of Tours as a topic for research. During the early stages of my work, I benefited from a World Universities Network fellowship at the Centre for Medieval Studies in Bergen, where Lars Boje Mortensen kindly acted as my advisor. Charles Burnett, Carolyn Muessig, Gillian Clark, Marcus Bull, Peter Cramer, Frank Coulson, Erika Kihlman, Denis Searby, Barbara Newman, Josh Smith, Thomas Haye, and the anonymous reader for Brill offered valuable comments on earlier versions of the work. My thanks are due to Wyndham Thomas, Michael Allen, Steve d'Evelyn, and Bill Flynn for advice on various points of detail, and to Ray Clemens for comments on my transcription of the Laudian manuscript discussed in Chapter 4. Colleagues at Northwestern University (in Medieval Studies, Classics, and the Alice Kaplan Institute for the Humanities) provided a friendly and stimulating environment in which much of the book was written. Among them, John Wynne never read any of what follows, but his conversation has been a sustaining reminder of what academic life really should be like. Finally, I would like to thanks my parents for their constant support and faith in me, and my wife and fellow medievalist, Cathy Hume, for her love, humour, and good sense. The dedication of this book to her is a small token of gratitude.

Abbreviations

Libraries

BL	British Library, London
BN	Bibliothèque nationale de France, Paris
Bodl.	Bodleian Library, Oxford

Publications

AHDLMA	*Archives d'histoire doctrinale et littéraire du Moyen Âge*
Cos.	Bernardus Silvestris, *Cosmographia*, ed. P. Dronke (Leiden: Brill, 1978)
CTC	*Catalogus Translationum et Commentariorum: Mediaeval and Renaissance Latin Translations and Commentaries*, ed. P. O. Kristeller, F. E. Cranz, V. Brown, J. Hankins, and R. A. Kaster (Washington, DC: Catholic University of America Press, 1960–)
CCSL	Corpus Christianorum, Series Latina
CCCM	Corpus Christianorum, Continuatio Mediaevalis
MGH	Monumenta Germaniae Historica
PL	*Patrologiae cursus completus, series latina*, ed. J.-P. Migne, 221 vols. (Paris: Migne, 1841–1864)

A Note on Texts and Translations

The edition used in this book is *Cosmographia*, ed. P. Dronke (Leiden: Brill, 1978), which is based on Bodl. Laud misc. 515, controlled by readings from a further seven manuscripts from a total of over fifty. Prose sections are cited by book, chapter, paragraph and line number (e.g. I.2.1.5), metres by book, chapter and verse (e.g. II.4.2). For some suggested improvements to Dronke's edition, see B. Löfstedt, "Notizen zur *Cosmographia* des Bernardus Silvestris", *Bulletin du Cange* 51 (1993): 203–8, and M. Minkova, "Bernardus Silvestris's *Cosmographia* and its Optimism (with an Appendix of Comments and Suggestions on the Text of the *Cosmographia*)", *The Journal of Medieval Latin* 13 (2003): 127–62. I have not made use of Vernet's unpublished critical edition of the *Cosmographia*, based on around fifty manuscripts, and announced in A. Vernet, "Bernardus Silvestris et sa *Cosmographia*", in *École Nationale des Chartes. Positions des thèses, 1937* (Nogent-le-Rotrou: Daupeley-Gouverneur, 1937), 167–74. A defective photocopy of this rare dissertation is available for consultation at the library of the Pontifical Institute of Mediaeval Studies in Toronto. I have translated quotations from the *Cosmographia* myself, with the occasional borrowing from Winthrop Wetherbee, where I thought his excellent translation could not be bettered. Unless otherwise specified, all other translations are my own.

Introduction

In March 1936 the English printer Guido Morris was busy in his garden shed. There, in rural Somerset, Morris was working on his beloved Columbian hand-press, type-setting a pamphlet to announce a new publishing venture: the Latin Press. As Morris explained in his announcement, the Press would produce reprints of major works of medieval Latin literature:

> Especially interested in mediaeval things, I have found—as every reader must—the lack of certain important works in readable editions, and some of these I hope to print as I have time for them. The books I shall generally chose will be those I have experienced the greatest difficulty in obtaining, or so far have only been able to read in the British Museum. Among those which I hope to print in the near future, may be mentioned the *Sic et Non* of Abaelard, and Bernardi Silvestris *De Mundi Universitate*.[1]

Morris was a genuine lover of Latin—as one would perhaps need to be to consider such a quixotic venture—and he is said to have bombarded a long-suffering friend with letters written in appalling Latin. His enthusiasm for the Middle Ages in particular was that of the Arts and Crafts medievalism of William Morris and Eric Gill. But Guido was more artist than entrepreneur, and the Latin Press, like many of his ambitious projects, never got off the ground.[2]

Guido Morris stands nonetheless as one of the more colourful modern admirers of the twelfth-century *Cosmographia* (as it is now known) of Bernard Silvestris. His appreciation for the work seems to have derived from reading the great populariser of Medieval Latin, Helen Waddell, whose 1933 novel *Peter Abelard* includes a memorable cameo of Bernard, "very young, but as wise as a troll", lecturing on creation to a rapt audience of students at the cathedral school of Tours.[3] But Bernard's allegory of the creation of the cosmos and man has also found admirers in more recent critical scholarship. There it has been

---

1 G. Morris, *The First Announcement of the Latin Press* (Langford, Som.: Latin Press, 1936), 1. I quote from the copy of this pamphlet in Bristol Central Library, SRC 8, B35130.

2 For the best account of Morris's career, see A. Baker, "The Quest for Guido", *The Private Library* 2 (1969): 139–76. I would like to thank Mr Baker for corresponding with me (in English) on this subject.

3 H. Waddell, *Peter Abelard* (London: Constable, 1933), 117–20. Morris in fact published just this short section of the novel under the title *Sylvestris Discourses* (Langford, Som.: Latin Press, 1936). See also Waddell's earlier eulogy of Bernard in *The Wandering Scholars* (London: Constable, 1927), 114–22.

 | DOI 10.1163/9789004268357_002

praised as "one of the greatest cosmological poems of world literature",[4] a work that "has earned the right to be compared with the *Divine Comedy*",[5] and one that "ranks among the finest products of twelfth-century humanism."[6]

The *Cosmographia* is indeed a remarkably inventive and engaging work. It is a creation story, but one with few obvious ressemblences to the orthodox Christian account. The dramatis personae are not the Father, Son and Spirit, but a family of creative goddesses who enter the scene as the world languishes at a low ebb in one of its periodic cycles of creation and destruction. Bernard's allegory envisages the reformation of the world as something of a feudal drama. The work opens with a rhetorical oration in which Natura accuses her mother Noys (the mind of God) of an abuse of sovereignty over primordial matter, Silva. The plight of Silva is painted in harrowing terms: she is a neglected subject, a grey-haired mother, forced to live in squalour and subject to random penetration by the forms. She comes now with her many children as a suppliant before God. Natura's plea is that Noys bestow upon Silva the beauty and splendour befitting this venerable and ancient being.

Natura's impassioned compalint had been in verse—specifically hexameters, as if to emphasise an epic dimension to Silva's plight. But now Noys replies in the measured prose of a mother reasoning with her quarrelsome child. Your request shall be granted, she says, but only because it comes at just that moment when God had decided to take Silva in hand. Noys then proceeds to bring order to matter through the harmonising proportions of the four elements. The world thus brought into being, Mundus, is married to soul, Endelechia, thereby animating creation. Parts of Silva, however, remain lurking in the cosmos, an ever-present threat to the happy marriage. Switching back to another verse form (elegiac couplets), Bernard now launches into a catalogue of over three hundred created beings, from the celestial hierarchy and fixed stars down to the flora and fauna of Earth. The first book then concludes with another section of prose, which returns from the concrete multiplicity of creation to the metaphysical discussion begun in the first prose chapter, apparently in imita-

4 R. Klibansky and F. Regen, *Die Handschriften der philosophischen Werke des Apuleius: ein Beitrag zur Überlieferungsgeschichte* (Göttingen: Vandenhoeck und Ruprecht, 1993), 47: "einer der großen kosmologischen Dichtungen der Weltliteratur."

5 J. C. Relihan, *Ancient Menippean Satire* (Baltimore, MD: Johns Hopkins University Press, 1993), 196. Contrast Gilson, cited below in Chapter 3, n. 15.

6 B. Stock, *Myth and Science in the Twelfth Century: A Study of Bernard Silvester* (Princeton, NJ: Princeton University Press, 1972), 37.

tion of the cycle of creation itself, which moves "from genus to species, from species to individual, and from individual back again to first principles."[7]

As this summary of just the first of its two books should make clear, the *Cosmographia* is a particularly complex and demanding work.[8] The aim of the present book is to offer a new reading of the treatise in its twelfth-century context. For whilst it has long been accepted that the *Cosmographia* was a major product of the vibrant intellectual life of twelfth-century Europe, there has been little consensus about where exactly in the period's intellectual landscape the work should be placed. Some have emphasised the theological aspects of the *Cosmographia*, seeing it as part of the 'pre-scholastic' background to the great syntheses of Christian doctrine that run from the *Sentences* of Peter Lombard in the 1150s through to the *summae theologicae* of the thirteenth century.[9] Early appraisals tended to be characterised by absolutist evaluations of the orthodoxy of the work,[10] but these assessments have developed into more nuanced discussion of Bernard's place in twelfth-century debate.[11] Others have focused on the literary or rhetorical features of the *Cosmographia*, reading it as an example of the revival of classicism associated with the 'twelfth-century renaissance', and a product of a scholarly milieu more interested in pagan poetry than Christian theology.[12] Literary concerns have also been prominent in criticism that has situated the *Cosmographia* in the history of allegory,[13] and

---

7 *Cos.* I.4.3.1–2: "de generibus ad species, de speciebus ad individua, de individuis ad sua rursus principia ..."

8 See Appendix I for a chapter-by-chapter synopsis of the entire work.

9 For this approach to twelfth-century thought, see A. M. Landgraf, *Introduction à l'histoire de la littérature théologique de la scolastique naissante*, ed. and trans. A.-M. Landry and L.-B. Geiger (Montréal: Institut d'Études Médiévales, 1973); and for the place of the *Cosmographia* in such a narrative, M. L. Colish, *Peter Lombard*, 2 vols. (Leiden: Brill, 1994), I: 301–97.

10 See especially Bernardus Silvestris, *De mundi universitate*, ed. C. S. Barach and J. Wrobel (Innsbruck: Wagner'schen Universitaets-Buchhandlung, 1876), xv–xviii; M. de Wulf, *Histoire de la philosophie médiévale*, 2nd ed. (Louvain: Institut Supérieur de Philosophie, 1905), 231–36; and É. Gilson, "La cosmogonie de Bernardus Silvestris", *AHDLMA* 3 (1928): 5–24.

11 T. Silverstein, "The Fabulous Cosmogony of Bernardus Silvestris", *Modern Philology* 46 (1948): 92–116; C. Ratkowitsch, *Die Cosmographia des Bernardus Silvestris: eine Theodizee* (Cologne: Böhlau, 1995).

12 C. H. Haskins, *The Renaissance of the Twelfth Century* (Cambridge, MA: Harvard University Press, 1927), 93–126; R. McKeon, "Poetry and Philosophy in the Twelfth Century: The Renaissance of Rhetoric", *Modern Philology* 43 (1946): 217–34 (222).

13 J. Whitman, *Allegory: The Dynamics of an Ancient and Medieval Technique* (Oxford: Clarendon Press, 1987), 218–60.

especially of representations of the goddess Natura,[14] in studies of the mixed form of prose and verse in which the work is written,[15] the development of technical manuals on the *ars poetriae*,[16] the distinctive philosophical poetic of the northern French schools,[17] and the development of Latin erotic lyric.[18] A third major critical perspective may be identified in those studies that have read the *Cosmographia* as a witness to the vitality and diversity of twelfth-century natural science prior to the reception of Latin translations of Aristotle's *libri naturales*.[19] Such studies have focused on Bernard's handling of his Latin philosophical sources,[20] or on the nature of his borrowings from recently translated Arabic scientific texts,[21] or on both, as in Brian Stock's magisterial study of the *Cosmographia*.[22]

---

14 E. R. Curtius, *European Literature and the Latin Middle Ages*, trans. W. R. Trask (New York: Pantheon, 1952), 108–13; G. D. Economou, *The Goddess Natura in Medieval Literature* (Cambridge, MA: Harvard University Press, 1972), 58–72; P. Dronke, "Bernard Silvestris, Natura, and Personification", *Journal of the Warburg and Courtauld Institutes* 43 (1980): 16–31; B. Newman, *God and the Goddesses: Vision, Poetry, and Belief in the Middle Ages* (Philadelphia: University of Pennsylvania Press, 2003), 51–65.

15 B. Pabst, *Prosimetrum: Tradition und Wandel einer Literaturform zwischen Spätantike und Spätmittelalter* (Cologne: Böhlau, 1994), I: 446–76; B. K. Balint, *Ordering Chaos: The Self and the Cosmos in Twelfth-Century Latin Prosimetrum* (Leiden: Brill, 2009); P. Dronke, *Verse with Prose from Petronius to Dante: The Art and Scope of the Mixed Form* (Cambridge, MA: Harvard University Press, 1994), 46–47.

16 D. Kelly, *The Arts of Poetry and Prose* (Turnhout: Brepols, 1991), 57–64.

17 W. Wetherbee, *Platonism and Poetry in the Twelfth Century: The Literary Influence of the School of Chartres* (Princeton, NJ: Princeton University Press, 1972), 158–86.

18 T. C. Moser, Jr., *A Cosmos of Desire: The Medieval Latin Erotic Lyric in English Manuscripts* (Ann Arbor: University of Michigan Press, 2004).

19 For the origins of the revisionist thesis of a twelfth-century 'discovery of nature', see M.-D. Chenu, "Découverte de la nature et philosophie de l'homme à l'École de Chartres au xiie siècle", *Cahiers d'histoire mondiale* 2 (1954): 313–25; and T. Gregory, *Anima mundi: la filosofia di Guglielmo di Conches e la scuola di Chartres* (Florence: Sansoni, 1955), 175–246. For further bibliography, see below, Chapter 2, n. 1.

20 M. F. McCrimmon, "The Classical Philosophical Sources of the *De mundi universitate* of Bernard Silvestris" (Doctoral Dissertation: Yale University, 1952); F. Tauste Alcocer, *Opus naturae: la influencia de la tradición del "Timeo" en la "Cosmographia" de Bernardo Silvestre* (Barcelona: Promociones y Publicaciones Universitarias, 1995).

21 R. Lemay, *Abu Ma'shar and Latin Aristotelianism in the Twelfth Century: The Recovery of Aristotle's Natural Philosophy through Arabic Astrology* (Beirut: American University, 1962), 258–84.

22 Stock, *Myth and Science*. For similar emphases, see F. Bezner, *Vela Veritatis: Hermeneutik, Wissen und Sprache in der 'Intellectual History' des 12. Jahrhunderts* (Leiden: Brill, 2005), 415–70.

Each of these critical perspectives has contributed significantly to our understanding of the *Cosmographia*, and it would be wrong to suggest that all the above-mentioned studies have been restricted to just one angle of interpretation. However, it is the particular argument of the present book that the position of the *Cosmographia* in the intellectual culture of the twelfth century will be best understood through an integrated account of its literary, theological and scientific themes. More specifically, the case will be made that Bernard Silvestris was drawing on certain literary techniques and traditions to create a distinctive response to some of the natural scientific and theological problems that occupied the minds of other contemporary masters in the northern French schools.

Underpinning this reading of the *Cosmographia* are the two concepts invoked by the title of this book: authority and imitation. Although these English words derive from Latin terms (*auctoritas* and *imitatio*), they now carry a set of connotations that is rather more negative than those of their medieval ancestors. Since the Reformation, the notion of authority has gradually come to be seen as antonymous with freedom of thought and expression, ideals by which modernity has often defined itself in opposition to its medieval past.[23] On the other hand, cultural unease with the notion of imitation has a more recent history. Although philosophical criticism of imitation as falsification can be traced all the way back to Plato's *Republic*, it was only in the later eighteenth century that the practice of literary imitation began to meet with sustained criticism in light of new Romantic ideals of originality, self-expression and individuality.[24] Both concepts, however, were central to the intellectual culture in which Bernard Silvestris worked, and in order to introduce the arguments of this book it will be helpful to look at each of them in some more detail.

In simplest terms, the *auctoritas* of a text in the Middle Ages most often derived from its antiquity, which was thought to be an indicator of its truthfulness.[25] Medieval scholars did not think they were discovering new facts about the world, but rather rediscovering ancient wisdom. There was

---

23 For the origins of such thinking, see G. R. Evans, *Problems of Authority in the Reformation Debates* (Cambridge: Cambridge University Press, 1992), 80–85.

24 G. Else and H. R. Elam, "Imitation", in *The New Princeton Encyclopedia of Poetry and Poetics*, ed. A. Preminger and T. Brogan (Princeton, NJ: Princeton University Press, 1983), 575–79.

25 A. J. Minnis, *Medieval Theory of Authorship: Scholastic Literary Attitudes in the Later Middle Ages* (London: Scolar Press, 1984), 10–15; and see in general J. Ziolkowski, "Cultures of Authority in the Long Twelfth Century", *Journal of English and Germanic Philology* 108 (2009): 421–48.

a sense that in a golden age of remote antiquity philosophy had been completed—*complere* is the verb used by Roger Bacon—and that only parts had been transmitted to the present.[26] The rediscovery of ancient knowledge took the form either of the physical recovery of previously unknown ancient texts, as in the great translation movement (*c.* 1075–1250) through which Bernard Silvestris lived, or else of a progressively deeper understanding of established sources from antiquity. Academic inquiry was therefore conducted primarily through the format of commentary, whose task it was to elucidate the knowledge latent in ancient sources. This was of course paradigmatically the case with Scripture, which could be seen as the divinely inspired source of all knowledge—even scientific knowledge, for Wisdom had "ordered all things in measure, and number, and weight" (Ws. 11:21).[27] But a similar reverence for the *auctoritas* of ancient sources also applied to the Church Fathers, as well as to ancient writers on the liberal arts (Donatus and Priscian for grammar, Cicero for rhetoric, Aristotle, Porphyry and Boethius for logic, and so on), whose relation to Scripture was characterised by propaedeutic metaphors of ascent (*gradus*) or service (*ancillae*).[28]

Among the liberal arts *auctores*, certain works were held in such high regard that they were thought to contain digests of universal learning. This approach is already apparent in the influential ninth-century commentators Johannes Scotus Eriugena and Remigius of Auxerre, whose allegorical interpretations

26 Roger Bacon, *Opus majus* II.14, ed. J. H. Bridges, 3 vols. 2nd ed. (London: Williams and Norgate, 1900), 67 (refering to Noah and his sons): "... quibus Deus dedit vivere sexcentos annos ad minus propter studium sapientiae complendum." Another well-known image was Boethius's description of Lady Philosophy's tattered dress, which symbolises the fragmentation of philosophical knowledge in the work of her so-called followers, especially the Stoics and Epicureans. See *De consolatione Philosophiae* I, pr. 1.4–5, ed. L. Bieler, CCSL 94 (Turnhout: Brepols, 1957): "Harum in extremo margine Π Graecum, in supremo uero Θ legebatur intextum atque in utrasque litteras in scalarum modum gradus quidam insigniti uidebantur, quibus ab inferiore ad superius elementum esset ascensus. Eandem tamen uestem uiolentorum quorundam sciderant manus et particulas quas quisque potuit abstulerant."

27 See below, Chapter 3, 95–96.

28 E. Peters, "Transgressing the Limits Set by the Fathers: Authority and Impious Exegesis in Medieval Thought", in *Christendom and its Discontents: Exclusion, Persecution and Rebellion, 1000–1500*, ed. S. L. Waugh and P. D. Diel (Cambridge: Cambridge University Press, 1996), 338–62; M.-T. d'Alverny, "La Sagesse et ses sept filles: recherches sur les allégories de la Philosophie et des arts libéraux du ixe au xiie siècle", in *Mélanges dediés à la mémoire de Felix Grat*, 2 vols. (Paris: Pecqueur-Grat, 1946–1949), I: 245–78 (repr. in d'Alverny, *Études sur le symbolisme de la Sagesse et sur l'iconographie*, ed. C. Burnett (Aldershot: Ashgate, 1993), I).

of Martianus Capella and Boethius uncovered the great depths of learning concealed beneath the surface meaning of the text.[29] And such encyclopaedic literary interpretation reached the peak of its popularity in the mid-twelfth century, around the time (*c.* 1159) that John of Salisbury was writing that ancients such as Vergil and Lucan had so adorned their works with learning that "the finished work [*opus consummatum*] would seem in some way to be an image of all the arts."[30] In such cases, the task of the commentator was not to proceed in what one might call a 'superficial', modern manner, using inter- and intra-textual references to elucidate the meaning of the work in question. Certainly there were some general observations of this kind, but these were typically confined to the academic prologues (*accessus*) to commentaries that would have constituted the first lecture in a master's exposition of a text. Thus in the *accessus* to his *c.* 1120 commentary on Boethius's *De consolatione Philosophiae*, William of Conches wrote that the *utilitas* of the work lay in its teaching that one should not rejoice excessively in prosperity nor be depressed by adversity, and that the author imitated the mixed form of Martianus Capella for the ethical purpose of offering the twin consolations of reason (prose) and delight (verse).[31]

But after such introductory comments, the twelfth-century commentator's characteristic approach was not 'superficial' but 'deep': he would focus on a particular phrase, and—often independent of immediate context—plumb the depths of its hidden moral or scientific meaning. Thus in William's glosses on

---

29 For the ancient origins and medieval development of this tradition of exegesis, see R. Lamberton, *Homer the Theologian: Neoplatonist Allegorical Reading and the Growth of the Epic Tradition* (Berkeley: University of California Press, 1986); and J. Chance, *Medieval Mythography: From Roman North Africa to the School of Chartres, A.D. 433–1177* (Gainesville: University of Florida Press, 1994).

30 John of Salisbury, *Metalogicon* I.24, ed. J. B. Hall and K. S. B. Keats-Rohan, CCCM 98 (Turnhout: Brepols, 1991), 52.27–33: "Illi enim per diacrisim quam nos illustrationem sive picturationem possumus appellare, cum rudem materiam historiae aut argumenti aut fabulae aliamve quamlibet suscepissent, eam tanta disciplinarum copia, et tanta compositionis et condimenti gratia excolebant, ut opus consummatum omnium artium quodam modo videretur imago."

31 William of Conches, *Glosae super Boetium*, ed. L. Nauta, CCCM 158 (Turnhout: Brepols, 1999), 5–6. Such considerations of *utilitas* and *cui parti philosophiae supponitur* were two of the seven standard topics discussed in what Hunt called the 'Type C Prologue', the most common in the twelfth century. See R. W. Hunt, "The Introductions to the *Artes* in the Twelfth Century", in *Studia Mediaevalia in Honorem Reverendi Patris Raymundi Josephi Martin* (Bruges: De Tempel, 1948), 85–112 (repr. in Hunt, *The History of Grammar in the Middle Ages: Collected Papers*, ed. G. L. Bursill–Hall (Amsterdam: John Benjamins, 1980), 117–44); and Minnis, *Medieval Theory of Authorship*, 18–28.

the first book of the *Consolatio*, a section of the work that is not lacking in philosophical or indeed dramatic interest, one of William's longest glosses concerns what to a modern reader might seem the relatively trivial remark that before his fall Boethius "used to scrutinise the radiance of the frosty moon."[32] The phrase *gelidae lunae* leads William into a lengthy discussion of how one of the planets, which are generally regarded as being bright and therefore hot, might legitimately be called "frosty". His suggestions are that either the cold of the moon is not inherent but something felt in its effects on Earth, or else that the moon might be considered cold due to the slowness of its orbit in relation to the other planets.

Because it was a frequent presupposition that ancient sources conveyed the truth, commentators were inclined to read back into these sources some of the different propositions that they, in a later age, believed to be true. Thus in his commentary on Cicero's *Somnium Scipionis*, the fifth-century philosopher Macrobius declared that Plato was "the sanctuary of truth itself", but much of Macrobius's understanding of Plato is refracted through his reading of the third-century AD philosopher Porphyry, whose Neoplatonic doctrines developed Plato's original ideas.[33] Where we might see such attitudes to the *auctoritas* of Plato as anachronistic, Macrobius and subsequent commentators would rather have seen evidence of the gradual uncovering of latent wisdom. Problems with this approach, however, could arise either when the *auctores* did not appear to agree or when the commentator's own ideas began to diverge too widely from his source. Thus, to return to William of Conches's treatment of Boethius's *gelida luna*, the commentator was aware that his thesis about the moon's relative slowness appeared to contradict the opinion of Calcidius and Macrobius that all the planets moved at the same speed. But by means of some "hermeneutic manoeuvring", as Lodi Nauta has called it, William argued that it was possible to preserve the authority (*auctoritatem servare*) of Calcidius and Macrobius by saying that a planet circles more slowly when it has less air below it to move.[34]

---

32 Boethius, *De consolatione Philosophiae* I, m. 2.9, ed. Bieler, 3: "uisebat gelidae sidera lunae"; William of Conches, *Glosae super Boetium*, ed. Nauta, 49–51.

33 Macrobius, *Commentarii in somnium Scipionis* I.6.23, ed. J. Willis (Leipzig: Teubner, 1963), 22.13–14: "Item scimus secundum Platonem, id est secundum ipsius veritatis arcanum ..." For discussion of sources, see Macrobius, *Commentary on the Dream of Scipio*, trans. W. H. Stahl (New York: Columbia University Press, 1952), 23–39.

34 William of Conches, *Glosae super Boetium*, ed. Nauta, xlix. For more on the relationship between interpretation and independent speculation, see L. Nauta, "The 'Glosa' as Instrument for the Development of Natural Philosophy: William of Conches's Commentary on Boethius", in *Boethius in the Middle Ages: Latin and Vernacular Traditions*

Discussions of *auctoritas* became more frequent in the twelfth century due to factors such as the polemics of the Investiture controversy, the proliferation of written documents in secular and ecclesiastical administration, and a level of academic exchange and debate that was becoming more sustained in the increasingly urbanised setting of the cathedral schools than it had been in the monasteries of the earlier Middle Ages. These discussions of *auctoritas* were by no means limited to natural scientific problems. In theology and law, for example, Peter Abelard, in his *Sic et non*, and Gratian, in his *Decretum*, were organising the complex, and sometimes apparently contradictory, traditions of early medieval biblical commentary and ecclesiastical legislation into systematic treatises. In the case of natural philosophy, the prospects of systematisation were complicated by the wealth of new scientific learning that was flowing into the northern French schools in the early twelfth century. Rediscovered scientific texts from Latin antiquity included Seneca's *Quaestiones naturales*,[35] Cicero's *De natura deorum*, the second book of which contained a valuable treatment of Stoic physics,[36] and the corpus of late Roman philosophical works attributed to Apuleius.[37] In addition to such sources, there was also, from the early twelfth century, the stimulus of a greatly increased influx of Latin translations of Arabic science from the peripheries of Christendom.[38] These new texts transmitted a range of medical and astrological learning that greatly expanded the mathematical framework of the early medieval quadrivium.

---

of the "Consolatio Philosophiae", ed. M. J. F. M. Hoenen and L. Nauta (Leiden: Brill, 1997), 3–39.

35 *Texts and Transmission: A Survey of the Latin Classics*, ed. L. D. Reynolds (Oxford: Clarendon Press, 1983), 376–78; M. Lapidge, "The Stoic Inheritance", in *A History of Twelfth-Century Western Philosophy*, ed. P. Dronke (Cambridge: Cambridge University Press, 1988), 81–112 (102–3).

36 *Texts and Transmission*, ed. Reynolds, 124–28; Lapidge, "Stoic Inheritance", 102.

37 See Klibansky and Regen, *Handschriften*; C. Burnett, "The Establishment of Medieval Hermeticism", in *The Medieval World*, ed. P. Lineham and J. L. Nelson (London: Routledge, 2001), 113–20; K. H. Dannenfeldt, "Hermetica philosophica", CTC, I: 137–64, with supplements in CTC, II: 423–24 and III: 425–26; and P. Lucentini, "L'Asclepius ermetico nel secolo xii", in *From Athens to Chartres: Neoplatonism and Medieval Thought. Studies in Honour of Edouard Jeauneau*, ed. H. J. Westra (Leiden: Brill, 1992), 397–420.

38 The fundamental work on the scientific translations by C. H. Haskins, *Studies in the History of Mediaeval Science* (Cambridge, MA: Harvard University Press, 1924), and the chapter in L. Thorndike, *A History of Magic and Experimental Science*, 8 vols. (New York: Columbia University Press, 1923–1958), II: 66–93, have been updated by the survey in M.-T. d'Alverny, "Translations and Translators", in *Renaissance and Renewal in the Twelfth Century*, ed. R. L. Benson and G. Constable, with the assistance of C. D. Lanham (Oxford: Clarendon Press, 1982), 421–62.

At a time when established curricular authors were being supplemented and even superseded by new philosophical sources, one approach—which in some ways ran parallel to the new genres in theology and canon law—was to write not a commentary but rather an independent treatise that styled itself as a natural scientific aid to the *lectio philosophorum*. Such was the *Dragmaticon philosophiae* (1144–1149) by William of Conches, which incorporates both new learning and independent speculation, and whose subject matter is described by its author as "useful for the pursuit of science and necessary for reading the philosophers."[39]

Another possibility was to compose a work that was not a commentary, but rather an *imitatio* of the ancients. Like *auctoritas, imitatio* was an important concept in the academic milieu in which Bernard Silvestris worked. As a category of thought, *imitatio* was not restricted to literature, but encompassed the theological ideal of *imitatio Christi* and the practice of students imitating the behaviour of charismatic schoolmasters.[40] In literary terms, imitation was a central component of an educational system in which pupils were to internalise the stylistic virtues of the *auctores* through memorisation and imitative composition. In the schoolroom, it could consist of such "introductory exercises in imitating prose and poetry" as those carried out by the pupils of Bernard of Chartres.[41] But since Roman antiquity, *imitatio* also had a creative, even a competitive dimension, since it originally involved transferring Greek models into Latin, and an expectation that Latin authors should aim to outdo their illustrious predecessors.[42] This idea of imitation as competition

39 William of Conches, *Dragmaticon philosophiae* I.1.7, ed. I. Ronca (with the Catalan *Summa de philosophia vulgari*, ed. L. Badia and J. Pujol), CCCM 152 (Turnhout: Brepols, 1997), 7.73–74: "ad scientiam utile et ad lectionem philosophorum necessarium".

40 D. de Rentis, *Die Zeit der Nachfolge: zur Interdependenz von 'imitatio Christi' und 'imitatio auctorum' im 12.–16. Jahrhundert* (Tübingen: Max Niemeyer, 1996); S. Jaeger, *The Envy of Angels: Cathedral Schools and Social Ideals in Medieval Europe, 950–1200* (Philadelphia: University of Pennsylvania Press, 1994).

41 See John of Salisbury, *Metalogicon* I. 24, ed. Hall and Keats-Rohan, 53.76–85; M. J. Carruthers, *The Book of Memory: A Study of Memory in Medieval Culture* (Cambridge: Cambridge University Press, 1990), 189–220; and for a case study of the wider influence of classroom exercises in rhetorical *descriptio*, D. Kelly, *The Conspiracy of Allusion: Description, Rewriting, and Authorship from Macrobius to Medieval Romance* (Leiden: Brill, 1999), 79–120.

42 See D. A. Russell, "*De imitatione*", in *Creative Imitation and Latin Literature*, ed. D. West and T. Woodman (Cambridge: Cambridge University Press, 1979), 1–16; and J. Ziolkowski, "The Highest Form of Compliment: *Imitatio* in Medieval Latin Culture", in *Poetry and Philosophy in the Middle Ages: A Festschrift for Peter Dronke*, ed. J. Marenbon (Leiden: Brill, 2001), 293–307 (300–301), who discusses Quintilian, *Institutio oratoria* X.2.1–28, on imitation and rivalry.

continued in the Middle Ages, and can be seen in the work of Charlemagne's biographer Einhard, whose *Vita Karoli* imitated both the structure and phrasing of Suetonius's *Augustus* in order to emphasise the historical stature of Charlemagne and his empire. The literary forms and figures transmitted by the ancients (such as the mixed prose and verse form and the goddess Natura) were not inert materials that medieval writers cut and pasted into new configurations; rather they were living ideas to be creatively adapted to particular situations.[43] "Good imitation", as Karl Morrison puts it, "must be continual invention."[44] And the best medieval imitators wove texts in which the warp of their own invention was highlighted by the weft of the *auctores* on whom they drew.

Imitation, then, was a literary technique particularly well suited to the circumstances of the twelfth-century schools, in which new texts and philosophical outlooks were explored alongside commentary on the traditional *auctores*. Of the general process whereby commentary on ancient texts could shade into fresh composition, Robert Lamberton has written: "The text generates a literature that stands in greater and greater opposition to the text's apparent meaning until at last a new text emerges, able to replace both."[45] And as an ancient paradigm of this process, David Dawson has pointed to the opening of the Gospel of John, which draws on Alexandrian allegorical exegesis of the first verses of Genesis (as the creation of the intelligible as opposed to the material universe) to propose that the light and darkness of Genesis are contained in the divine *logos*. As Dawson writes of John:

> This interpreter wants his audience to hear in the new myth the echoes of these verses from Genesis because he wants his audience to realize that the earlier story in Genesis is not really the separate, independent story it might appear to be; instead, it is merely part of a new, more comprehensive story.[46]

---

43 See P. Dronke, *Poetic Individuality in the Middle Ages: New Departures in Poetry, 1000–1150*, 2nd ed. (London: Westfield College, 1986), xi–xvi, 11–22, for criticism of the *Toposforschung* approach of Curtius's *European Literature*.

44 K. F. Morrison, *The Mimetic Tradition of Reform in the West* (Princeton, NJ: Princeton University Press, 1982), ix: "... as it was, for example, when Bach imitated Vivaldi."

45 Lamberton, *Homer the Theologian*, 286.

46 D. Dawson, *Allegorical Readers and Cultural Revision in Ancient Alexandria* (Berkeley: University of California Press, 1992), 130.

What the Evangelist gives us, in a narrative refashioned by a tradition of exegesis, is "Genesis properly understood, that is, Genesis as absorbed by a larger story".[47]

A good example of a twelfth-century philosophical work that absorbs traditional material into "a larger story" is provided by the *De essentiis* by Hermann of Carinthia. Completed in 1143 in Béziers, between the French schools to the north, where Hermann had studied, and the Iberian peninsula to the south, where he had worked as a translator from Arabic, the *De essentiis* imitates aspects of Plato's *Timaeus*—especially the opening dialogue sections and Plato's dramatic description of God constructing the universe—and it incorporates numerous Platonic doctrines, but these Hermann supplemented with a range of new Arabic science.[48] Through *imitatio*, the *auctoritas* of Plato is preserved, just as the commentator creates space to introduce new material that will update—or should one say elucidate?—the teachings of his Platonic model.[49]

The *Cosmographia* of Bernard Silvestris was probably completed by 1147 or 1148, the dates of the French sojourn of Pope Eugenius III, who is reported by a gloss in one of the manuscripts to have heard the work recited *in Gallia*.[50] Like Hermann's work from the same decade, the *Cosmographia* is a sophisticated *imitatio*, one that addressed current philosophical problems by weaving contemporary science into a mythological narrative similar to some of the school authorities by which it was inspired. The first chapter of the present book explores the school milieu in which Bernard was working. Moving away from

47 Dawson, *Allegorical Readers*, 131.

48 Hermann of Carinthia, *De essentiis*, ed. and trans. C. Burnett (Leiden: Brill, 1982), 16–17, 25–43. See also the *Anticlaudianus* of Alan of Lille, "in many respects a Christian version of the late fifth-century *Marriage of Philology and Mercury*, in which the pagan Martianus Capella had described how the seven liberal arts can elevate the soul to be fit for heaven." (*Medieval Literary Theory and Criticism c. 1100–c. 1375: The Commentary Tradition*, ed. A. J. Minnis and A. B. Scott, with the assistance of D. Wallace (Oxford: Clarendon Press, 1988), 124.)

49 Lamberton, *Homer the Theologian*, 286, rightly observes that the imitator is not necessarily a self-conscious innovator: "[T]he author conceives his relationship to a tradition as one of imitation, or at least of participation, but his own historicity (given concrete form in the traditions under discussion by the commentaries that mediate between him and the text) has transformed that text and imposed on the imitation a radically new structure of meaning."

50 Bodl., Laud misc. 515, f. 188v (glossing *Cos.* I.3.55, which mentions Eugenius): "Iste Eugenius fuit papa in cuius presencia liber iste fuit recitatus in Gallia, et captat eius benivolenciam." For further discussion of this event, see below, Chapter 3, n. 7.

the traditional association of Bernard with 'the school of Chartres', this chapter locates Bernard in the schools of a different French city, Tours, at which he is actually known to have taught. Twelfth-century Tours had a grand tradition of scholarship stretching back to the Carolingian revival of learning, when Alcuin of York, the city's most famous abbot, helped establish the study of some of the liberal arts *auctores* that Bernard imitated and revised. This chapter pays particular attention to one liberal arts manuscript from twelfth-century Tours, for it contains a number of contemporary poems that engage in different ways with ancient sources, and these offer an immediate intellectual backdrop to the practices of imitation and revision at work in the *Cosmographia*.

The central two chapters of the book turn to the *Cosmographia* itself. Each examines a major topic of twelfth-century scholarship to which Bernard's treatise was conceived as a contribution. Chapter 2 is concerned with the relationship of the *Cosmographia* to the renewed interest in the rational analysis of the natural world, which was stimulated by translations of Greek and Arabic science. Where other contemporaries such as William of Conches and Hermann of Carinthia were absorbed by the rational investigation of phenomena in physiology, meteorology and astronomy, we shall see that Bernard was the philosopher *par excellence* of the human possibilities and limitations associated with the study of nature. His *Cosmographia* explored the human dimensions of natural science through the mixed form employed by Martianus Capella and Boethius in their late antique accounts of the educational formation of the human soul. This chapter looks at the ways in which Bernard imitated and adapted these *auctores*, shifting his emphasis away from the seven liberal arts of early medieval scholarship to incorporate the more pragmatic science of twelfth-century medicine and astrology.

Chapter 3 examines the *Cosmographia*'s contribution to another major concern in the twelfth-century schools, namely the relationship of natural science to the study of Scripture and theology. For just as this period saw a developing interest in the study of nature, so too there was a growing unease about naturalistic accounts of God's creation. Were there really waters located above the firmament, as the Book of Genesis maintained? Was Eve literally created from Adam's rib? And should the third person of the Trinity, the Holy Spirit, be understood as identical with the world soul of the Platonists? Such questions were being posed by the contemporary masters Peter Abelard and William of Conches, both of whom were attacked by monastic critics in the 1140s. This was also the decade of the *Cosmographia*'s composition, and the question of the relationship of Bernard Silvestris to these debates has been one of the major interests of the criticism on his work. The present study will argue that Bernard's position on this theological question must once again be understood in terms of his imitation and adaptation of ancient models. For where others masters were using the critical concept of a 'covering' (*integumentum*) to unearth controversial theological doctrines beneath the surface of pagan authors, Bernard turned the *integumentum* from a technique of interpretation into a principle of composition. He stepped back from the direct equation of pagan philosophy with Christian theology to produce a new composition in imitation of the ancients, whose inherent difficulty was valued as a way of leading the mind to consider various possibilities of interpretation and thereby draw nearer to the truth.

Chapter 4 concludes with a survey of the reception of the *Cosmographia* in France and England during the century and a half after its publication. It was considered the mark of a good literary *imitatio* that it in turn was deemed worthy of imitation, and along with a small number of other twelfth-century works, the *Cosmographia* achieved a level of *auctoritas* second only to that of the ancients themselves. One index of Bernard's *auctoritas* is the substantial volume of glosses that accrued around the text of the *Cosmographia* in this period. These are studied in this chapter for the first time. Yet Bernard's work was a contribution to a very particular moment of twelfth-century intellectual history, and as that moment passed so his unifying solution to a cluster of problems that had been occupying the minds of northern French scholars in the 1140s began to lose its relevance. To study the reception of the *Cosmographia*, then, is to trace the changing academic attitudes of the period, as schools merged into universities, as the *auctores* were adapted to more pragmatic, pastoral needs, and as the theological problems that Bernard had treated so allusively became subject to increasingly formalised procedures of scholastic analysis.

It is part of the special appeal of the *Cosmographia* that it is both a beautiful and demanding work of literature and also a document of primary importance for the intellectual history of its period. Balanced between the early and later Middle Ages, the *Cosmographia* represents a culmination of the liberal arts scholarship of the Carolingian schools; but it also anticipates some of the new approaches in theory and practice taken by natural philosophers in the later medieval universities. It is the aim of this book to convey something of the intellectual appeal and the historical significance of this many-sided work.

CHAPTER 1

Bernard and the Schools of Tours

Who was the author of the *Cosmographia*? To some early critics Bernard Silvestris was none other than Bernard of Chartres, and therefore the master at the centre of a great school of philosophical studies that flourished at Chartres Cathedral in the eleventh and twelfth centuries.[1] Both these propositions have to different degrees proved suspect. That Bernard of Chartres and Bernard Silvestris were distinct figures, the first dying at least a decade before the second wrote, was demonstrated as early as 1893.[2] The second claim, that a number of the most significant philosophers of twelfth-century France were masters at Chartres, is one that R. W. Southern subjected to largely convincing criticism in a series of studies from the 1970s to the 1990s.[3] However, one consequence of Southern's work has, ironically, been a reaffirmation of the 'Chartrian' status of Bernard Silvestris. For some have argued that though not all the masters once linked to Chartres necessarily taught there, 'the school of Chartres' remains nonetheless a convenient label for "a body of ideas and the scholars and poets who developed them."[4] Thus although there is no evidence

1 See, for example, *Histoire littéraire de la France; ouvrage commencé par des religieux bénédictins de la Congrégation de Saint-Maur, et continué par des membres de l'Institut* (*Académie des inscriptions et belles-lettres*), 32 vols. (Paris: Imprimerie nationale, 1835–1981), XII (1869): 264; Bernardus Silvestris, *De mundi universitate*, ed. Barach and Wrobel, vii–xiv; and R. L. Poole, *Illustrations of the History of Medieval Thought in the Departments of Theology and Ecclesiastical Politics* (London: Williams and Norgate, 1884), 117–18, who corrected the error in the 1920 edition (p. 148).

2 B. Hauréau, "Maître Bernard", *Bibliothèque de l'École des Chartes* 54 (1893): 792–94; R. Merlet and J. A. Clerval, *Un manuscrit chartrain du xie siècle* (Chartres: Garnier, 1893); J. A. Clerval, *Les écoles de Chartres au moyen-age, du ve au xvie siècle* (Chartres: Selleret, 1895), 158–62. For the case in favour of the identity of the two Bernards, see C.-V. Langlois, "Questions d'histoire littéraire: Maître Bernard", *Bibliothèque de l'École des Chartes* 54 (1893): 225–50 (237–47).

3 R. W. Southern, "Humanism and the School of Chartres", in his *Medieval Humanism and Other Studies* (Oxford: Blackwell, 1970), 61–85; *Platonism, Scholastic Method and the School of Chartres* (Reading: University of Reading, 1979); "The Schools of Paris and the School of Chartres", in *Renaissance and Renewal*, ed. Benson *et al.*, 113–37; *Scholastic Humanism and the Unification of Europe*, 2 vols. (Oxford: Blackwell, 1995–2001), I: 58–101.

4 Wetherbee, *Platonism and Poetry*, xii. For criticism of this "new Chartrian hypothesis", see Southern, *Scholastic Humanism*, I: 93–100, and for a judicious review of the debate, J. Marenbon, "Humanism, Scholasticism and the School of Chartres", *International Journal of the Classical Tradition* 6 (2000): 569–77.

 | DOI 10.1163/9789004268357_003

that Bernard Silvestris either studied or taught at Chartres, this redefinition of the school of Chartres has made it possible to speak of him as "a Chartrian poet", and even as "[le] premier des grands maîtres chartrains".[5]

There is no doubt that Bernard's thought does share important affinities with contemporary thinkers whose relationship to the school at Chartres remains to varying degrees controversial, in particular with William of Conches, John of Salisbury, and Thierry of Chartres, to whom the *Cosmographia* was dedicated. Such affinities will be given due place in this study. However, the emphasis on the so-called Chartrian character of Bernard's thought has tended to deflect attention from the city where he is in fact known to have taught, namely Tours. It is well known, for example, that Bernard's student, the poet Matthew of Vendôme, wrote: "I was taught the art of composition by the glory of Tours, Master Silvestris, the jewel of studies and ornament of the school."[6] The problem is that the schools of Tours remain something of an enigma. At a time when there was much discussion of the masters and schools that were driving the vibrant intellectual scene in northern France, Tours is rarely mentioned. The city does not, for example, appear in the accounts of contemporary scholarship to be found in Peter Abelard's *Historia calamitatum*, John of Salisbury's *Metalogicon*, William of Tyre's *Chronicon*, or in the anonymous *Metamorphosis Goliae*. And yet, as Peter Dronke has rightly insisted, Tours could lay claim to an impressive—and sometimes controversial—succession of scholars from the eleventh and twelfth centuries, including Berengar of Tours, Hildebert of Lavardin, Roscelin of Compiègne, Adelard of Bath, and Bernard himself.[7]

It would be anachronistic to expect a distinctive intellectual programme to emerge from the writings of these scholars.[8] In a period before the regularisa-

5 Wetherbee, *Platonism and Poetry*, 22; Baudri of Bourgueil, *Poèmes*, ed. and trans. J.-Y. Tilliette, 2 vols. (Paris: Les Belles Lettres, 1998–2002), I: xxxvii. B. Bonn and A. M. Bautier, *Index Scriptorum Novus Mediae Latinitatis: Supplementum (1973–2005)* (Geneva: Droz, 2005), 44, still describes Bernard Silvestris as "chancellor of Chartres" (*Carnotensis cancellarius*).

6 *Epistola* I.3.69–70, in *Mathei Vindocinensis Opera*, ed. F. Munari, 3 vols. (Rome: Storia e letteratura, 1977–1988), II: 90: "Me docuit dictare decus Turonense magistri | Silvestris, studii gemma, scolaris honor." Matthew was also probably the author of an epitaph for Bernard in Angers, Bibliothèque municipale 303, discovered by C. Wollin, "Das Epitaphium Bernardi Siluestris in der Handschrift Angers B. M. 303", *Sacris Erudiri* 42 (2003): 369–401. The couplet (on f. 127v) reads: "Non Bernardus obit, sed abit; non interit, i<m>mo | Incipit; hospes erat, admodo civis erit." The *hospes* of the second line picks up on an important term from the *Cosmographia*: see below, Chapter 3, 116–17.

7 *Cos.*, 1–15.

8 However, the criticisms directed by P. Godman, "Ambiguity in the *Mathematicus* of Bernardus Silvestris", *Studi medievali*, 3rd ser. 31 (1990): 583–648 (587–91) against the whole notion of a 'school of Tours' (as developed by Dronke, in the previous note) seem unduly negative.

tion of higher learning in the thirteenth-century universities, masters and their students moved with relative freedom from school to school, and intellectual affinities were understood in terms that were less institutional than individual. What contemporary titles there were for 'schools of thought' commonly invoke adherence to a particular master (*Porretani*) or doctrine (*Nominales*) rather than a place of study.[9] Nonetheless, individual twelfth-century schools did have their own academic specialisms—Laon in biblical exegesis, Paris in logic and theology, Chartres in natural philosophy, Orléans in the study of the classics—and these have been well documented by modern scholarship.[10] By contrast, the schools of Tours have received scant attention.[11] Part of the problem has been a paucity of surviving evidence. Margaret Gibson remarked of the city's principal abbey that "[t]he loss of most of the primary documents has deterred modern scholars from making a study of St Martin's, Tours."[12] Sources do survive, however, and in sufficient numbers for scholarship to reconstruct important aspects of the city's early medieval past.[13] Although there is less evidence than for other contemporary schools, certain observations can still be made about the situation at Tours, and these are observations well worth making for the scattered light they shed on the otherwise obscure conditions in which one of the most interesting writers of the period was working. The aim of this chapter, then, is not to provide a comprehensive survey of surviving evidence relating to the masters, manuscripts and academic life of twelfth-century Tours. Rather its purpose is to paint a more impressionistic picture of the immediate intellectual landscape in which Bernard worked, emphasising

---

9 W. J. Courtenay, "Schools and Schools of Thought in the Twelfth Century", in *Mind Matters: Studies of Medieval and Early Modern Intellectual History in Honour of Marcia Colish*, ed. C. J. Nederman, N. van Deusen, and E. A. Matter (Turnhout: Brepols, 2010), 13–45.

10 For a sample, see V. I. Flint, "The 'School of Laon': A Reconsideration", *Recherches de théologie ancienne et médiévale* 43 (1976): 89–110; S. Ferruolo, *The Origins of the University: The Schools of Paris and their Critics, 1100–1215* (Stanford, CA: Stanford University Press, 1985); É. Jeauneau, *L'âge d'or des écoles de Chartres* (Chartres: Houvet, 1995); and R. H. Rouse, "Florilegia and Latin Classical Authors in Twelfth- and Thirteenth-Century Orléans", *Viator* 10 (1979): 131–60.

11 The most complete documentation remains that of E. Lesne, *Histoire de la propriété ecclésiastique en France, V: les écoles de la fin du viii*[e] *siècle à la fin du xii*[e] (Lille: Giard, 1940), 137–47.

12 M. Gibson, "Letters and Charters Relating to Berengar of Tours", in *Auctoritas und Ratio: Studien zu Berengar von Tours*, ed. P. Ganz, R. B. C. Huygens and F. Niewöhner (Wiesbaden: Harrassowitz, 1990), 5–23 (6, n. 6).

13 See S. A. Farmer, *Communities of Saint Martin: Legend and Ritual in Medieval Tours* (Ithaca, NY: Cornell University Press, 1991).

those details which, in one way or another, bring into relief the treatise at the centre of this study.

The City and Its Schools

There were three major religious communities in twelfth-century Tours. In the centre of the ancient city was the cathedral of Saint-Maurice, whose third-century patron was martyred with his Theban legion on the banks of the Rhone, an incident that is alluded to in Bernard's *Cosmographia*.[14] To the south-west of the city walls lay the abbey of Saint-Martin. The abbey housed the tomb of its patron, which due to Roman burial laws had to be located outside the walls of the Gallo-Roman *civitas*. With the development of the abbey of Saint-Martin as a pilgrimage site, the Châteaneuf district around the church grew in the early Middle Ages, and this area was itself enclosed by walls in 918. Finally, across the river on the northern bank of the Loire lay the Benedictine house of Marmoutier, founded by St Martin in around 372.[15] There is no indication that Bernard Silvestris was a Benedictine monk, and it seems likely, given both the presence of an early copy of his *Mathematicus* in a Saint-Martin manuscript (Tours, Bibliothèque municipale 300) and the more distinguished scholarly tradition of that abbey, that it was of the community of Saint-Martin, rather than the cathedral, that Bernard was a member.

Saint-Martin was a rich community. Its ninth-century abbot was the lord of 20,000 serfs, and it had extensive landholdings both in the Touraine and further afield—in Poitou, the Auvergne, and even in Italy—and the abbey's Romanesque basilica, rebuilt in 1014 after a fire, was a vast edifice that rivalled some of the great pilgrimage destinations of Europe.[16] The members of the community had resisted the general movement, legislated for by the Synod of Aachen (817), to establish the Rule of St Benedict as the standard observance in Carolingian monasteries. Instead, and in common with other prestigious communities, such as the Parisian abbeys of Saint-Denis and

14 *Cos.* I.3.257–58: "Influxit Rodanus, ubi nobile vidit Agauno | Certamen turbe martiris ante mori." The cathedral is now called Saint-Gatien.

15 C. Brühl, *Palatium und Civitas: Studien zur Profantopographie spätantiker Civitates vom 3. bis zum 13. Jahrhundert, I: Gallien*, 2 vols. (Cologne: Böhlau, 1975–1990), I: 101–110; H. Leclercq, "Tours", *Dictionnaire d'archéologie chrétienne et de liturgie*, 15 vols. (Paris: Letouzey et Ané, 1924–1953), XV: 2570–677.

16 E. R. Vaucelle, *La collégiale de Saint-Martin de Tours des origines à l'avènement des Valois (397–1328)* (Tours: Péricat, 1908), 104–46.

Saint-Maur-des-Fossés, Saint-Martin was allowed to become a house of canons—*canonicus* being a term that came into use in the early ninth century to distinguish from Benedictine houses those that observed a less strict monastic rule.[17] The life of the canons was a comfortable one. The *Ritual* written by the canon Péan Gatineau between 1226 and 1237 indicates that the chapter comprised up to 150 canons, and that each received a prebend of land and wine and a house in the cloister.[18] We know that Bernard Silvestris possessed other property in Tours because sometime between 1178 and 1184 his nephew, Gerbert Boceau, sold a house in the city left him by his uncle.[19] The aristocratic opulence of the chapter was criticised in around 1180 by the Benedictine monk Guibert of Gembloux, who complained that, instead of documenting the continuing miracles of St Martin, the canons were busy "playing games of chance, sporting after birds of the sky, and hunting hares, roes, and deer with their dogs."[20]

The city of St Martin is mentioned only once in the *Cosmographia*, appearing at the end of a catalogue of famous rivers:

> Secana prosiliit, ubi grandia germina regum—
> Pipinos, Karolos—bellica terra tulit.
> Emicuit Ligeris, ubi Martinopolis inter
> Sidereos fluvios pictaque rura sedet. (I.3.261–64)

> The Seine arose where warlike earth brought forth
> The mighty stock of Charles and Pippin—kings.
> Where Martin's city lies, 'twixt starry waters
> And the tinted fields, the Loire did gleam.

As in the forest and mountain catalogues of the *Cosmographia*, the rivers in this section of the poem are carefully ordered. Moving from east to west and from antiquity to the present, Bernard used his river catalogue to trace a history of *translatio imperii* from the Tigris and Euphrates, through the Jordan, the Simois and the Rhone, to reach the penultimate river, the Seine, whose *bellica terra* brought the transfer of power to Charlemagne, crowned emperor

17 O. G. Oexle, *Forschungen zu monastischen und geistlichen Gemeinschaften im westfränkischen Bereich* (Munich: Fink, 1978), 120–33.

18 Vaucelle, *La collégiale*, 193.

19 Vernet, "Bernardus Silvestris", 168. The Boceau family were major land-holders in the Touraine and prominent donors to monasteries in the area: see L. de Grandmaison, *Cartulaire de l'Archevêché de Tours*, 2 vols. (Tours: Péricat, 1892–1894), II: 315.

20 Farmer, *Communities of Saint Martin*, 200, n. 13.

in AD 800. As with Chrétien de Troyes' famous contention that *chevalrie* came to France hand-in-hand with *clergie*, Bernard's catalogue concludes with a quieter *locus amoenus*, a city whose greatest abbot, Alcuin of York, appointed by Charlemagne in 804, wrote that he wished France to become "a new Athens."[21] Bernard, in other words, inhabited a city of cultural significance, a modern *polis* benefiting from the transmission of learning that accompanied the transfer of empire to the Carolingian dynasty.

Alcuin was one of the architects of the educational reforms aimed at the *renovatio* of the Carolingian Church.[22] Part of those reforms involved the reassertion of a set of ideals for Christian education, found in Augustine and Cassiodorus, in which the liberal arts, based on the study of grammar and the Latin classics, were seen as essential training for the correct understanding of Scripture, and thus a foundation for orthodoxy throughout Charlemagne's empire.[23] Naturally, these reforms were also felt in Tours. Soon after his appoint-

---

21 *Ep.* 170 (to Charlemagne, 799), ed. E. Dümmler, *MGH: Epistolae IV* (Berlin: Weidmann, 1895), 279.20–22: "Nec fastidiosa segnities legentium benivolentiae magistri iuste deputari debebit; si, plurimis inclitum vestrae intentionis studium sequentibus, forsan Athenae nova perficeretur in Francia, immo multo excellentior." An association between Alcuin, Saint-Martin and *translatio studii* appears in the 884–887 chronicle of Notker Balbulus, *De gestis Karoli imperatoris* I.2 (*PL* 98, 1373B; quoted by É. Jeauneau, *Translatio studii. The Transmission of Learning: A Gilsonian Theme* (Toronto: PIMS, 1995), 13): "Dedit [Karolus] autem illi [*i.e.* Alcuino] abbatiam sancti Martini iuxta Turonicam ciuitatem, ut quando ipse absens esset, illic requiescere et ad se confluentes docere debet. Cuius in tantum doctrina fructificauit, ut moderni Galii siue Franci antiquis Romanis et Atheniensibus aequarentur." The story was repeated in the 1270s by John of Wales, *Compendiloquium* X.6 (in Johannes Gallensis, *Communiloquium sive summa collationum* (Venice: Georgius de Arrivabenis, 1496), f. 231rv): "Venit Alcuinus cum sociis ad eum [*i.e.* Karolum] in omnibus scripturis exercitatus, utpote Bede discipulus. Quem rex secum usque in finem retinuit nisi cum ad bella iret: deditque ei abbatiam beati Martini, ut quando rex absens quiesceret, et ad se confluentes doceret. Et sic Franci equati sunt Romanis."

22 R. McKitterick, *The Frankish Church and the Carolingian Reforms, 789–895* (London: Royal Historical Society, 1977), 45–79; A. Diem, "The Emergence of Monastic Schools: The Role of Alcuin", in *Alcuin of York: Scholar at the Carolingian Court*, ed. L. A. J. R. Houwen and A. A. MacDonald (Groningen: Egbert Forsten, 1998), 27–44; M. de Jong, "From *Scolastici* to *Scioli*: Alcuin and the Formation of an Intellectual Elite", in *Alcuin of York*, ed. Houwen and MacDonald, 45–57.

23 See the legislative *Epistola de litteris colendis*, ed. A. Boretius, *Capitularia Regum Francorum I, MGH: Leges II* (Hannover: Hahn, 1883), 78–79; and for Alcuin's contribution to this legislation, L. Wallach, *Alcuin and Charlemagne: Studies in Carolingian History and Literature* (Ithaca, NY: Cornell University Press, 1959), 204–12. See also L. Holtz, "Alcuin et la renaissance des arts libéraux", in *Karl der Grosse und sein Nachwirken*, 5 vols., ed. P. L. Butzer, M. Kerner and W. Oberschelp (Turnhout: Brepols, 1996–1998), I: 45–60; and

ment to Saint-Martin, Alcuin wrote to tell Charlemagne of how he had been feeding his community at Tours with the honey of sacred scripture, inebriating them with the liberal arts, and nourishing them with grammatical study.[24] The Carolingian revival of the liberal arts—and especially of the *ars grammatica*—as a foundation for Christian education had an enormous influence on the subjects, texts and procedures of learning adopted in European schools through to the twelfth century.[25] And in a sense, the legacy of Alcuin's insistence on the close connection between the liberal arts and scriptural exegesis was to create an institutional and conceptual space for later thinkers, such as Bernard Silvestris, to probe the borderlands of theology and philosophy.

A related aim of the Carolingian reforms was the production of accurate manuscript copies of liturgical, canon law and scriptural texts, along with the supporting patristic commentaries and propaedeutic authorities in the arts.[26] Here again, Tours was at the forefront of wider developments. In the same 796–797 letter to Charlemagne, Alcuin wrote of his plans to transplant British books (*flores Brittaniae*) from York to his paradise on the Loire, which would become a *hortus conclusus* where the Lord would one day come to gather his fruit.[27] From the time of Alcuin's abbacy, the already active Saint-Martin scriptorium began to produce a number of famous deluxe Bibles, and became a

A. Bücker, "Christianizing the Arts: From Augustine's *De ordine* to Carolingian Thought", in *Signs of Change: Transformations of Christian Traditions and their Representations in the Arts, 1000–2000*, ed. N. H. Petersen, C. Clüver and N. Bell (Amsterdam: Rodopi, 2004), 173–89.

24 *Ep.* 121 (to Charlemagne, 796–797), ed. Dümmler, 176.32–4: "Ego vero Flaccus vester secundum exhortationem et bonam voluntatem vestram aliis per tecta sancti Martini sanctarum mella scriptuarum ministrare satago; alios vetere antiquarum disciplinarum mero inaebriare studeo; alios grammaticae subtilitatis enutrire pomis incipiam."

25 M. Irvine, *The Making of Textual Culture: 'Grammatica' and Literary Theory, 350–1100* (Cambridge: Cambridge University Press, 1994), 313–33; W. Wetherbee, "The Study of Classical Authors: From Late Antiquity to the Twelfth Century", in *The Cambridge History of Literary Criticism, II: The Middle Ages*, ed. A. Minnis and I. Johnson (Cambridge: Cambridge University Press, 2005), 99–144 (112).

26 R. McKitterick, *The Carolingians and the Written Word* (Cambridge: Cambridge University Press, 1989), 165–210.

27 *Ep.* 121, ed. Dümmler, 177.6–13: "Ideo haec vestrae excellentiae dico, ne forte vestro placeat totius sapientiae desiderantissimo consilio, ut aliquos ex pueris nostris remittam, qui excipiant inde nobis necessaria quaeque et revehant in Frantiam flores Brittaniae; ut non sit tantummodo in Euborica hortus conclusus, sed in Turonica emissiones paradisi cum pomorum fructibus, ut veniens Auster perflaret hortos Ligeri fluminis et fluant aromata illius, et novissime fiat quod sequitur in cantico, unde hoc adsumpsi paradigma: 'Veniat dilectus meus in hortum suum, et comedat fructum pomorum suorum.'"

publishing house for other monasteries.[28] Sadly, very little of the abbey's medieval collection can be reconstructed. For the library of Saint-Martin, dispersed after the Revolution, had already been decimated by sackings of the abbey by sixteenth-century Huguenots and eighteenth-century Revolutionaries.[29] Nor is there a surviving medieval catalogue from Saint-Martin. However, there are a number of extant volumes containing a twelfth-century *ex libris*, and by adding these to a collated list of E. K. Rand's catalogue of manuscripts in "the script of Tours" (the distinctive form of Caroline miniscule developed by the abbey's scriptorium in the ninth century) and the earliest, pre-Revolution catalogue of the Saint-Martin collection, it is possible to identify some surviving codices that may have been in the twelfth-century collection.[30] The total number of volumes to meet these strict criteria is forty-one.[31] Clearly this represents a small fraction of the manuscript resources that would have been available to Bernard at Saint-Martin. Even in the ninth century, it is likely that the abbey's

---

28 For the history of the scriptorium, see E. K. Rand, *Studies in the Script of Tours*, 2 vols. (Cambridge, MA: Mediaeval Academy of America Press, 1929); L. W. Jones, "The Script of Tours in the Tenth Century", *Speculum* 14 (1939): 179–98; Jones, "The Art of Writing at Tours from 1000 to 1200 A.D.", *Speculum* 15 (1940): 286–98; B. Bischoff, "Aus Alkuins Erdentagen", in Bischoff, *Mittelalterliche Studien: ausgewählte Aufsätze zur Schriftkunde und Literaturgeschichte*, 3 vols. (Stuttgart: Hiersemann, 1966–1981), II: 12–19; D. Ganz, "Mass Production of Early Medieval Manuscripts: The Carolingian Bibles from Tours", in *The Early Medieval Bible: Its Production, Decoration and Use*, ed. R. Gameson (Cambridge: Cambridge University Press, 1994), 53–62; and P. Gasnault, "Le scriptorium de Saint-Martin à l'époque carolingienne", *Mémoires de la Société Archéologique de Touraine* 62 (1997): 73–86.

29 C. de Grandmaison, *Procès-verbal du pillage par les Huguenots des reliques et joyaux de Saint-Martin de Tours en mai et juin 1562* (Tours: Mame, 1863); L. Delisle, "Notice sur les manuscrits disparus de la Bibliothèque de Tours pendant la première moitié du xixe siècle", *Notices et extraits des manuscrits* 31 (1883): 157–356.

30 The Saint-Martin catalogue compiled by B. de Montfaucon (in *Bibliotheca Bibliothecarum Manuscriptorum Nova*, 2 vols. (Paris: Briasson, 1739), II: 1335–41) is reprinted in M. Collon's catalogue of the Bibliothèque municipale de Tours: *Catalogue général des manuscrits des bibliothèques publiques de France: départements, XXXVII*, 2 vols. (Paris: Plon-Nourrit, 1900–1905), II: 1063–69.

31 One volume, containing Cicero's *De senectute* and *Somnium Scipionis* with the commentary of Macrobius (BN nouv. acq. lat. 454) has been reattributed to the scriptorium of Corbie by B. Bischoff, "Hadoard und die Klassikerhandschriften aus Corbie", in Bischoff, *Mittelalterliche Studien*, I: 49–63 (59). It should also be noted that Rand's palaeographic study contains only "certain specimens of the later development of the script of Tours", that is after the tenth century (*Studies*, I: viii), so this selection of forty-one manuscripts will be unduly weighted to the earlier period.

collection rivalled those of other major Carolingian scriptoria such as Saint-Gall, Reichenau and Murbach, whose surviving catalogues indicate collections of between three and four hundred volumes each.[32] It is not possible, therefore, to use the manuscripts as the basis for a detailed study of the school of the kind that has been undertaken in the cases of Laon and Saint-Gall, let alone to chart Bernard's reading and responses to his sources through particular manuscripts, as has been possible in the case of William of Malmesbury.[33]

What then do these forty-one volumes contain? Unsurprisingly, the majority (twenty-eight) contain liturgical and biblical texts, patristic and early medieval theology, and saints' lives—the core of any ecclesiastical collection in the Middle Ages. In addition, a further three ninth-century manuscripts contain texts relating to the cult of St Martin (including the *Vita* by Sulpicius Severus) and the writings of the city's sixth-century bishop, Gregory of Tours.[34] The remaining ten volumes contain liberal arts texts, and are therefore more relevant to Bernard's intellectual concerns. Most of these books were copied in the ninth and tenth centuries. There are two surviving works of lexicography: a *Liber glossarum*[35] and the *De compendiosa doctrina* of Nonius Marcellus.[36] There are also copies of Suetonius,[37] Augustine's *De musica*,[38] and two compilations of quadrivial and computistical texts. One includes Boethius's *De arithmetica*, Aratus's *Phenomena* and Bede's *De natura rerum*.[39] The other contains the second book of Boethius's *De arithmetica* together with a glossed copy of

32 B. Bischoff, "Libraries and Schools in the Carolingian Revival of Learning", in Bischoff, *Manuscripts and Libraries in the Age of Charlemagne*, trans. and ed. M. Gorman (Cambridge: Cambridge University Press, 1994), 93–114.

33 J. J. Contreni, *The Cathedral School of Laon from 850 to 930: Its Manuscripts and Masters* (Munich: Arbeo-Gesellschaft, 1978); A. A. Grotans, *Reading in Medieval St. Gall* (Cambridge: Cambridge University Press, 2006); R. M. Thomson, *William of Malmesbury*, 2nd ed. (Woodbridge, Suff.: Boydell, 2003).

34 Tours, Bibliothèque municipale 1018 and 1019 (Monfaucon, nos. 155 and 154; Rand, I: 200–201); The Hague, Museum Meermanno-Westreenianum Q 3 + BN lat. 9733 (Monfaucon, no. 154; Rand, I: 178–79).

35 Tours, Bibliothèque municipale 850 (Monfaucon, no. 272; Rand, I: 187).

36 Leiden, Universiteitsbibliotheek Voss lat. F 73 (Rand, I: 105); with the twelfth-century *ex libris* (f. 1r): "iste liber est de armario beati martini turonensis."

37 BN lat. 6115; with the same twelfth-century *ex libris* as in n. 36.

38 Tours, Bibliothèque municipale 286 (Monfaucon, no. 29; Rand, I: 99–100).

39 Tours, Bibliothèque municipale 334 + BN nouv. acq. lat. 1612–1614 (Monfaucon, no. 42; Rand, I: 185–86).

De consolatione Philosophiae, prefaced by the treatise of Lupus of Ferrières on the metres of the *Consolatio*.[40]

If one were to assume that Saint-Martin retained exemplars of all the manuscripts produced by the scriptorium, then one could add to the collection copies of Vergil's *opera*, with the commentaries of Servius and Tiberius Claudius Donatus,[41] and (of more relevance to the *Mathematicus*), texts of Livy and Lucan.[42] Naturally, though, this would still only account for a fraction both of the *auctores* used by Bernard and of the liberal arts resources of Saint-Martin, the cathedral, and the abbey of Marmoutier, a short walk over the eleventh-century bridge that crossed the Loire. For all these losses, there is at least one manuscript that offers an insight into the twelfth-century school at Saint-Martin. Tours, Bibliothèque municipale 300 is a small schoolbook produced in the Saint-Martin scriptorium during the second half of the twelfth century.[43] The manuscript consists of two originally separate parts, the second of which (ff. 49r–75r) contains a number of texts that relate to the literary and philosophical interests of the twelfth-century schools, and provides a contemporary context in which to read the *Cosmographia*. There is, for instance, a copy of Boethius's *De trinitate* (ff. 49r–67r), which begins with a discussion of the division of sciences that is important for understanding the conception of natural philosophy in the *Cosmographia*.[44] That Boethius's treatise was very much in vogue in the twelfth-century schools is indicated by the extensive interlinear and marginal glossing and the fragmentary commentary (on ff. 67r–70v) that follows it by Thierry of Chartres, the dedicatee of the *Cosmographia*.[45] The final four folios of the volume take us more directly to the poetic interests of the school. They include four poems that adapt ancient sources in a way that illuminates Bernard's handling of his own source material: these will be treated in more detail in the final section of this chapter.

Turning from manuscripts to masters and students, it would seem that Saint-Martin followed a wider European cycle of academic vibrancy in the

40 Tours, Bibliothèque municipale 803 (Monfaucon, no. 2; Rand, I: 193–94). The marginal and interlinear glosses on the *Consolatio* have so far apparently been overlooked by students of the commentary tradition on the work.

41 Rand, I: 90–92, 127–28, 141–43.

42 Rand, I: 95–97, 130, 176–77, 191.

43 Montafaucon, no. 67; Rand, I: 202.

44 See below, Chapter 2, 52–53.

45 See *Commentaries on Boethius by Thierry of Chartres and his School*, ed. N. M. Häring (Toronto: PIMS, 1971), 26, where the Tours manuscript is described as having "the best text" of Thierry's *Commentum*. The glosses on *De trinitate* itself (ff. 49r–67r) do not appear to have been studied.

ninth cetury, followed by a phase of decreased activity—during which Saint-Martin, like other abbeys, had to contend with Viking raids—before a period of obvious intellectual renewal from the mid-eleventh century. In the abbacy of Alcuin (796–804) the school drew students from far beyond the Loire valley, including Rabanus Maurus (from Mainz), whom a ninth-century manuscript depicts being presented to St Martin by Alcuin.[46] Later in the ninth century, St Odo of Cluny (*c.* 878–942) is reported to have studied Priscian and Vergil at Saint-Martin, until a vision of a beautiful vase filled with snakes warned him of the deceptive charms of pagan authors.[47] From there the next major figure associated with the school would be Berengar of Tours, who taught liberal arts at Saint-Martin between around 1025 and 1040. It is possible that during this period Lanfranc of Bec, Berengar's future adversary in the Eucharistic controversy, heard Berengar lecture at Tours on his journey from Pavia to Normandy.[48] Although Berengar's denial of the material transformation of the bread and wine in the Eucharist received ecclesiastical condemnation on a number of occasions, including at the Council of Tours (1054), he was fondly remembered in the Touraine, receiving a favourable mention in a Saint-Martin chronicle and being honoured with an annual memorial service in the church.[49] Another controversial philosopher, Roscelin of Compiègne, was a canon at Saint-Martin between around 1100 and 1110, and his time in the city may have overlapped with that of the Englishman Adelard of Bath, whose prose and verse *De eodem et diverso* reports Adelard's liberal arts studies in the city around the turn of the twelfth century, including discussion of the universals debate in which Roscelin played a prominent part.[50] A final important precursor to Bernard at Tours was Hildebert of Lavardin, Archbishop of Tours (1125–1133), likewise the

46 For the international reach of the school, see *Gallia Christiana XIV*, ed. B. Hauréau (Paris: Coignard, 1856), 155–56; the manuscript image is in Vienna, Österreichische Nationalbibliothek 652, f. 2v.

47 Johannes Italus, *Vita S. Odonis* I (*PL* 133, 48D–49A).

48 M. T. Gibson, *Lanfranc of Bec* (Oxford: Clarendon Press, 1978), 15–23, 199; Gibson, "Letters and Charters", 10.

49 See *Cos.*, 8–9; and for the chronicle (written by a certain Peter, who is attested as *magister scolarum* at Saint-Martin in the 1140s and 1150s), see *Recueil de chroniques de Touraine*, ed. A. Salmon (Tours: Ladevèze, 1854), 57): "Fuit hoc tempore Beringarius Turonensis, amator pauperum." The editor notes that a Vatican manuscript substitutes for "Beringarius" the words "Beatus Ingarius", thereby creating a new saint—"dans le seul but de détruire un témoinage contemporain favorable au célèbre hérésiarque Tourangeau".

50 C. J. Mews, "Nominalism and Theology before Abaelard: New Light on Roscelin of Compiègne", *Vivarium* 30 (1992): 4–33; Adelard of Bath, *Conversations with his Nephew: "On the Same and the Different", "Questions of Natural Science", and "On Birds"*, ed. and

author of a work that combined prose and verse, the *De querimonia carnis et spiritus*, whose theme is alluded to in the *Cosmographia*.[51]

The place of Bernard Silvestris in the school is somewhat obscure. His name is not to be found in the admittedly patchy record of surviving Saint-Martin charters, which were occasionally attested by the *magister scolarum*.[52] Besides the *Cosmographia*, a number of extant works have been attributed to Bernard that may illuminate the workings of the school in the second quarter of the twelfth century. The problem is that Bernard's name came to carry a certain authority among later generations, and all the other treatises ascribed to him in manuscripts are to varying degrees dubious. The work for which Bernard was probably best known in the Middle Ages is in fact a brief letter on household management, the *Epistola de cura rei familiaris*, which survives in hundreds of manuscripts, including vernacular translations. The letter is addressed from a certain "Bernardus", and later copyists often took this to be Bernard Silvestris (perhaps because he was a known authority on the *ars dictaminis*) or else Bernard of Clairvaux. There has been no detailed study of the Latin text, but recent scholarship has tentatively proposed a thirteenth-century date for the letter.[53] Also of doubtful authenticity is a divinatory compilation called the *Experimentarius*, of which Charles Burnett has argued that only a small but significant part (the preface) may be Bernard's work.[54]

---

trans. C. Burnett, with the collaboration of I. Ronca, P. Mantas España and B. van den Abeele (Cambridge: Cambridge University Press, 1998), 22–23, 36–37.

51 *Cos.* II.4.7–12 (italics added): "Velle dei, mixtura modum, modulatio nexum, | Nexus amititiam pariat sacer, | Ne pigeat mentem cecas habitare tenebras | Hospitiumque pati grave corporis, | Nec propria *de carne* queat fecisse *querelam* | *Spiritus* imperiis subiectior."

52 E. Mabille, "La Pancarte Noire de Saint-Martin de Tours, brulée en 1793, et restituée d'après les textes imprimés et manuscrits", *Mémoires de la Société Archéologique de Touraine* 17 (1865): 319–540 (348–55). The plural form (*magister scolarum*) usually employed in the attestations did not in general denote different buildings but rather different instructional classes (M. M. Hildebrandt, *The External School in Carolingian Society* (Leiden: Brill, 1992), 72, 102–3), which in the case of Saint-Martin are known to have included the *scolae* of reading, writing and chant (E. Lesne, *Histoire de la propriété ecclésiastique en France, IV: les livres, 'scriptoria', et bibliothèques du commencement du viii*[e] *à la fin du xi*[e] *siècle* (Lille: Giard, 1938), 339).

53 R. Miguel Franco, "El texto de la *Epistola de cura rei familiaris* en el *Compendium Morale* de Geremia da Montagnone", *Rivista de Literatura Romanica* 20 (2008): 81–99 (81–82). The edition printed in C. D. M. Cossar, *The German Translations of the Pseudo-Bernhardine "Epistola de cura rei familiaris"* (Göppingen: Alfred Kümmerle, 1975), 97–102, is a collation of the text in *PL* 182, 647A–651A, and two of the Latin manuscripts.

54 C. Burnett, "What is the *Experimentarius* of Bernardus Silvestris? A Preliminary Survey of the Material", *AHDLMA* 44 (1977): 79–125 (repr. in Burnett, *Magic and Divination in the*

More controversial have been the two incomplete commentaries on Vergil's *Aeneid* and Martianus Capella. Scholars have been divided on the authorship of these works, with a clear majority (7-4 in the criticism I have read) expressing scepticism about a Bernardine attribution.[55] The external evidence for Bernard's authorship of the *Aeneid* commentary consists of an attribution in a late fifteenth-century manuscript—one of the five copies now known—and an entry of dubious value in the 1412 library catalogue of Amplonius of Ratinck, which refers to a "commentum Bernhardi Silvestris super 6 libros Eneidos".[56] Internal evidence strongly suggests that the author of the *Aeneid* commentary also wrote an unattributed commentary on Martianus Capella preserved in one English manuscript from *c.* 1200–1220.[57] The fullest case for Bernard's authorship of these works has been presented by Peter Dronke.[58] However, his argument that John of Salisbury refers (at *Policraticus* VIII.24) to the unnamed author of an *Aeneid* commentary in terms similar to those in which he describes (at *Policraticus* III.8) the author of another work probably by Bernard, the *Mathematicus*, is weakened by the fact that John's summary of the commentary in fact differs in many details from the one now attributed to Bernard.[59] Given the distinct genres of the *Cosmographia* and the commentaries, it is hard to make a convincing case for Bernard's authorship on

Middle Ages: Texts and Techniques in the Islamic and Christian Worlds (Ashgate: Aldershot, 1996), XVII). See further below, Chapter 3, n. 121.

55 For the extensive literature, see D. L. Pike, "Bernard Silvestris' Descent into the Classics: The *Commentum super sex libros Aeneidos*", *International Journal of the Classical Tradition* 4 (1998): 343–63, to which should be added M. Evans, "The *Ysagoge in Theologiam* and the Commentaries Attributed to Bernard Silvestris", *Journal of the Warburg and Courtauld Institutes* 54 (1991): 1–42, and S. Gersh, "(Pseudo-?) Bernard Silvestris and the Revival of Neoplatonic Virgilian Exegesis", in *Sophies maietores: Chercheurs de sagesse. Hommage à Jean Pépin*, ed. M.-O. Goulet-Cazé, G. Madec, and D. O'Brien (Paris: Institut d'Études Augustiniennes, 1992), 573–93.

56 The commentary is attributed to Bernard Silvestris in BN lat. 16246, f. 44r: "Incipit Commentum Bernardi Silvestris super sex Libros Eneidos Virgilii." For the text of Amplonius's catalogue, see P. Lehmann, *Mittelalterliche Bibliothekskataloge Deutschlands und der Schweiz* (Munich: Beck, 1918–1939), II: 5–95, where the volume in question (no. 35, pp. 15–16) is said also to contain the otherwise unattested "commentum Bernhardi Silvestris super Theodolum".

57 É. Jeauneau, "Note sur l'École de Chartres", *Studi medievali*, 3rd ser. 5 (1964): 821–65 (844–46).

58 P. Dronke, "Bernardo Silvestre", in *Enciclopedia Virgiliana*, ed. F. della Corte, 5 vols. (Rome: Istituto della Enciclopedia Italiana, 1984–1990), I: 497–500.

59 J. W. Jones, Jr., "The So-Called Silvestris Commentary on the *Aeneid* and Two Other Interpretations", *Speculum* 64 (1989): 838–48.

internal grounds alone: the doctrinal parallels adduced by Dronke are somewhat general, and can be countered by points of divergence whose relative weight is hard to assess.[60] It may be significant that the earliest French manuscript of the *Aeneid* commentary, BN lat. 3804 A, which was copied in the early thirteenth century, comes from Saint-Martial de Limoges, where one of the earlier French *Cosmographia* manuscripts (BN lat. 2770) was also copied. On the other hand, both commentaries have marked affinities with two English texts, the *Ysagoge in theologiam* and *Plato ad ostendendum*.[61] Should this be taken as evidence of English authorship of the commentaries, or rather of the early influence of Bernard's works across the Channel, of a piece with the enthusiastic English reception of his *Cosmographia* that we shall encounter in Chapter 4 of this book? Given the lack of persuasive evidence on either side, I do not think the commentaries may at present be attributed to Bernard with any great certainty.

Neither of the commentaries circulated with other texts from the schools of Tours. But this is not the case with the remaining works associated with Bernard, a group of four poems based on ancient declamations. We shall return to the content of these declamatory poems later in the chapter, but let us first consider their authorship. The three shorter works—*De gemellis* (86 lines), *De paupere ingrato* (32 lines), and *De nato Ethiope* (38 lines)—are all anonymous, but have been labelled 'school of Bernard' or 'school of Tours' on account of their thematic similarities to, and occurrence within manuscripts that contain both the *Cosmographia* and the longest of the declamatory poems, an 854-line astrological fable known as the *Mathematicus*.[62] The *Mathematicus* is attributed to Bernard in only three of its nineteen manuscripts—one each from the

60 Stock, *Myth and Science*, 36–37, n. 42; T. Silverstein, "Review of *The Commentary on the First Six Books of the Aeneid of Vergil Commonly Attributed to Bernardus Silvestris*, ed. J. W. Jones and E. F. Jones (Lincoln, Nebraska: University of Nebraska Press, 1977)", *Speculum* 54 (1979): 154–57.

61 C. Baswell, "The Medieval Allegorization of the *Aeneid*: MS Cambridge, Peterhouse 158", *Traditio* 41 (1985): 181–237; Evans, "The *Ysagoge in Theologiam*".

62 *Cos.*, 6–7. *De gemellis* (*incipit*: "Roma duos habuit—res est non fabula vana") and *De paupere ingrato* (*incipit*: "Mesta parens misere paupertas anxietatis") have been edited by R. E. Edwards, "Poetic Invention and the Medieval *Causae*", *Mediaeval Studies* 55 (1993): 183–217. *De nato Ethiope* (*incipit*: "Protrahit in vicium levitas sexum muliebrem") was edited by A. Vernet, "Poésies latines des xiie et xiiie siècles (Auxerre 243)", in *Mélanges dédiés à la mémoire de Félix Grat*, 2 vols. (Paris: Pecqueur-Grat, 1946–1949), II: 251–75 (258–59). There have been two recent editions of the *Mathematicus*: D. M. Stone, "Bernardus Silvestris, *Mathematicus*: Edition and Translation", *AHDLMA* 63 (1996): 209–83; and *Mathematicus*, ed. and trans. J. Prelog, M. Heim and M. Kiesslich (St Ottilien: EOS, 1993).

thirteenth, fourteenth and fifteenth centuries.[63] This relatively weak support for a Bernardine attribution is buttressed by Gervase of Melkley's *Ars poetica* (1208–1216), which names Bernard as the author, and by an unknown reader in Pontigny around the turn of the thirteenth century, who added Bernard's name to a *Policraticus* manuscript at the point where John of Salisbury quotes an unattributed couplet from the *Mathematicus*.[64]

However, Christine Ratkowitsch has called into question Bernard's authorship of the *Mathematicus* by drawing attention to certain differences between its prosody and that of the *Cosmographia*.[65] Comparing the first work with the *Cosmographia*'s six chapters in elegiac couplets—also the metre of the *Mathematicus*—Ratkowitsch has discovered clear variations in the use of enjambment and caesuras in the two works. Ratkowitsch's statistics bring valuable precision to a feeling no doubt shared by many readers that the poetry of the *Mathematicus* is quite different from that of the *Cosmographia*. But does it follow that one author was not responsible for both works, or could it rather be that Bernard was skilled enough to adapt his prosody to the varying demands of didactic and narrative verse? By way of comparison, we may note that the disparity observed by Ratkowitsch between the use of trithemimeral caesuras in the elegiac hexameters of the *Cosmographia* (36%) and the *Mathematicus* (27%) is in fact slightly less marked than that between Ovid's *Fasti* (35%) and his *Ibis* (25%).[66] Just as this stylometric evidence has not been used to challenge the Ovidian attribution of either work, it is not clear that it undermines the case for Bernard's authorship of the *Mathematicus*.

---

63 See Cambrai, Bibliothèque municipale 977, f. 156r [s. xiii]: "Bernardus"; Vatican City, Biblioteca Apostolica Vaticana Reg. lat. 370, f. 226v [s. xiv]: "Incipit mathematicus Bernardi Silvestris"; and BN lat. 11345 [AD 1455], a florilegium that has the marginal note "b sil" next to *Mathematicus* excerpts on ff. 15v, 29r, 32v, 42r, and 66r. From what has been reconstructed of the work's stemma (*Mathematicus*, ed. Prelog *et al.*, 25–30), these manuscripts do not appear to have closely related texts, a fact that would strengthen their attributions. For her edition Stone takes as her base the text in the Saint-Martin schoolbook, Tours, Bibliothèque municipale 300, ff. 72r–73v, for it is "likely to be closest to the lost original" (233).

64 Gervase of Melkley, *Ars poetica*, ed. H.-J. Gräbener (Münster: Aschendorff, 1965), 37, 64, 80; *Policraticus* III.8, ed. Keats-Rohan, 195.115–18: "Quod et quidam temporis nostri scriptor egregius, infidelium tamen uerbis, eleganter expressit: 'Ridiculos hominum uersat sors caeca labores; | saecula nostra iocus ludibriumque deis'" (*Math.* 175–76)—with the annotation in Montpellier, Faculté de médecine 60, f. 37r: "Bernardus Silvestris".

65 C. Ratkowitsch, "Astrologie und Selbstmord im Mathematicus: zu einem Gedicht aus dem Umkreis des Bernardus Silvestris", *Wiener Studien* 112 (1999): 175–229 (179–80).

66 J. P. Poe, *Caesurae in the Hexameter Line of Latin Elegiac Verse* (Wiesbaden: Steiner, 1974), Table Ia [94–95].

Furthermore, Ratkowitsch does not mention Peter of Blois (1125x30–1215), whose writings quote two proverbial couplets written by his *magister*, both of which come from the *Mathematicus*.[67] Peter is thought to have studied at Tours in the 1140s or 1150s, before traveling to Bologna to continue his education: he instructed his nephew to send him "the poems and trifles I wrote in Tours", which are presumably those "youthful love poems and adolescent trifles" that Peter piously declined to send to a later correspondent.[68] On the basis of his *Mathematicus* quotations, it has been assumed that Peter's master at Tours was Bernard Silvestris. However, Bernard is nowhere referred to by name, and another mention of a master in Peter's letters—"as a child when I was studying the art of versification, I applied myself, on the advice of my master, not to fables but to the truth of historical narratives"—could just as well, if not more probably, refer to another teacher, namely Peter the chronicler, who is attested as *magister scolarum* at Saint-Martin in the 1140s and 1150s.[69]

---

67 See *Compendium in Iob* (*PL* 207, 800D): "Sicut magister meus elegantissime dicit: 'Denigrat meritum dantis mora factaque raptim | Munera plus laudis plusque favoris habent'" = *Math.* 673–74; and *Ep.* 1, to Henry II (*PL* 207, 1B): "Sicut satis urbane scribit quidam magister meus: 'Est orare ducum species violenta jubendi | Et quasi nudato supplicat ense potens'" = *Math.* 717–18. A third quotation from the *Mathematicus* is conflated with line 66 of Hildebert of Lavardin's *Vita beate Marie Egiptiace* (ed. N. K. Larsen, CCCM 209 (Turnhout: Brepols, 2004), 235) in *Ep.* 51, to John of Salisbury (*PL* 207, 156A): "Porro si cogitatis vestris promissionibus stare, nec mihi accepta, nec vobis honesta est ista dilatio, testimonio poetae: 'Dilatio magna ruina est | *Et mora donandi non leve crimen habet*'" = *Math.* 696 (in italics).

68 *Ep.* 12 (*PL* 207, 39B): "Mitte mihi versus et ludicra, quae feci Turonis"; *Ep.* 62 (*PL* 207, 172C): "Quod autem amatoria juventutis, et adolescentiae vestrae ludicra postulas ad solatium taediorum, consiliorum non arbitror, cum talia tentationes excitare soleant, et fovere." These compositions have not been identified, but they may be the group of nine amatory and mythological lyrics (known as 'Collection X') that appear in various combinations in seven manuscripts, and which have been dated by their earliest codex to around 1150. See R. W. Lenzen, "Überlieferungsgeschichtliche und Verfasseruntersuchungen zur lateinischen Liebesdichtung Frankreichs im Hochmittelalter" (Doctoral Dissertation: University of Bonn, 1973), 87–97; and *Petri Blesensis Carmina*, ed. C. Wollin, CCCM 128 (Turnhout: Brepols, 1998), 124–29, 585–631. One of these lyrics, *De terre gremio* (ed. Wollin, 601–5, with commentary) looks to have been influenced by the *Cosmographia*. R. W. Southern, "The Necessity for Two Peters of Blois", in *Intellectual Life in the Middle Ages: Essays Presented to Margaret Gibson*, ed. L. Smith and B. Ward (London: Hambledon, 1992), 103–18, expressed well-founded scepticism about the attribution of secular lyrics to Peter of Blois, but his reservations need not apply to this early phase of Peter's writing career, before he turned to more serious subjects (as described in the following note).

69 *Ep.* 101 (*PL* 207, 314A): "Scio mihi plurimum profuisse, quod cum in arte versificatoria parvulus erudirer, praecipiente magistro mihi materiam non de fabulis, sed de historiarum veritate sumebam." On Master Peter, see above, n. 49.

It is conceivable then that neither Bernard Silvestris nor Master Peter was the *magister* to whom Peter of Blois ascribed the fable of the *Mathematicus*. But if so, we may have to posit the existence of a third, otherwise unknown master at Tours—one who shared Bernard's interests in fable and astrology, and also wrote poetry of widely recognised literary merit—and over such a figure would surely hang the blade of Ockam's razor.

Ars dictaminis

In addition to these writings with more or less reliable manuscript attributions, Bernard may also have written an as yet unidentified work on the theory of letter-writing (*ars dictaminis*). We know from one of Matthew of Vendôme's verse epistles that Bernard instructed him in this art at Tours:

> Me docuit dictare decus Turonense magistri
> Silvestris, studii gemma, scolaris honor.
> Dictando didici quid scribat amicus amico,
> Subiectus domino, mancipioque potens.[70]

> Master Silvestris—glory of Tours, learning's gem,
> And the pride of the school—taught me *dictamen*.
> I learned thereby how a friend should address a friend,
> A subject his lord, and a master his man.

It is not surprising that the school at Saint-Martin developed a speciality in the teaching of letter-writing. With its extensive lands, the community had long been interested in improving the production standards of charters, which, with their salutations and subscriptions, were considered a form of letter. From the influential collection of model acts in the eighth-century Tours Formulary, and the diplomatic work of Fridugisus of Tours—Alcuin's successor as abbot, chancellor at the court of Louis the Pious from 819, and the author of an imperial formulary as well as the philosophical treatise *De nihilo et tenebris*—the abbey of Saint-Martin was at the forefront of efforts to refine the juridical formulas and literary phrasing of official documents.[71] The production of charters in the

---

70 *Epistola* I.3.69–72, in *Mathei Vindocinensis Opera*, ed. Munari, II: 90.

71 P. Gasnault, "Les actes privées de l'abbaye de Saint-Martin de Tours du viie au xiie siècle", *Bibliothèque de l'École des Chartes* 112 (1954): 24–66 (29–32); R.-H. Bautier, "Chancellerie et culture au Moyen Âge", in *Cancelleria e cultura nel Medio Evo*, ed. G. Gualdo (Vatican City: Archivo Segreto Vaticano, 1990), 1–75 (6–9); P. Depreux, "La tradition manuscrite des

scriptorium was closely associated with the workings of the school; as in other communities, the head of the school was often also the chancellor.[72] Indeed in the 1140s, the decade of the *Cosmographia*'s composition, one of the *magistri scolarum* attested at Saint-Martin, a certain Josbertus, was probably responsible for compiling one of the substantial *pancartes* that was burned during the Revolution.[73]

The rules of charter composition were sometimes treated by manuals on the *ars dictaminis*, such the anonymous *Aurea gemma Gallica* (1153–1155), the oldest of the surviving French *artes*, and one whose origins may lie in the schools of Tours.[74] As a solemn record of the possessions of the community, the charter was a document deemed worthy of rhetorical ornamentation. This often occurred in the preamble (*arenga*), which would lay out a religious rationale for the charter.[75] A good example can be found in a Saint-Martin charter from around 1111, which records the resolution of a land dispute between Helias, Count of Le Mans, and the chapter of Saint-Martin, and is witnessed by Hildebert, then still Bishop of Le Mans, Fulcherius, the *magister scolarum* of Saint-Martin, and Roscelin of Compiègne. After beginning with the statement that God endowed man with reason to comprehend things and a mind to retain reason's deliberations, the charter continues:

> Sed quoniam, intervenientibus multimodis hujus saeculi perturbationibus, vel ipsa quam nemo subterfugit morte, per solam hominum memoriam nihil poterat aeternaliter retineri, volens genus humanum aliqua in tertiam quartamve progeniem, seu perpetuam custodire memoriam, per

'Formules de Tours' et la diffusion des modèles d'actes aux viiie et ixe siècles", *Annales de Bretagne et des Pays de l'Ouest* 111 (2004): 55–71.

72 Clerval, *Les écoles de Chartres*, 17, 22, 30, 47; T. Webber, *Scribes and Scholars at Salisbury Cathedral, c. 1075–c. 1125* (Oxford: Clarendon Press, 1992), 83.

73 Mabille, "La Pancarte Noire", 335. A *pancarte* recorded transactions that had not previously been committed to writing; a "cartulary, in contrast to a pancarte, copied together transactions which had already been preserved in writing and sealed" (C. B. Bouchard, "Monastic Cartularies: Organizing Eternity", in *Charters, Cartularies, and Archives: The Preservation and Transmission of Documents in the Medieval West*, ed. A. J. Kosto and A. Winroth (Toronto: PIMS, 2002), 22–32 (26)).

74 F. J. Worstbrock, M. Klaes and J. Lütten, *Repertorium der Artes dictandi des Mittelalters, I: von den Anfängen bis um 1200* (Munich: Wilhelm Fink, 1992), 90–92; *Petri Blesensis Carmina*, ed. C. Wollin, CCCM 128 (Turnhout: Brepols, 1998), 119–22.

75 H. Fichtenau, *Arenga: Spätantike und Mittelalter im Spiegel von Urkundenformeln* (Graz: Böhlau, 1957), esp. 131–32 for the themes of memory and written records treated in the charter below.

> sola litterarum monimenta omnia quae dicuntur, vel fiunt, ab oblivionis interitu cognovit posse defendi.[76]
>
> But because—due to the many kinds of intervening confusion in this world, or that death which no man evades—nothing was able to be retained forever in the mind by human memory alone, the human race, wanting to preserve some things unto the third or fourth generation or in perpetual memory, recognised that only through memorials of writing could all that is said and done be kept from the destruction of oblivion.

This is a classically constructed period, which contains allusions both to the Bible ("in tertiam quartamve progeniem"; *cf.* Exodus 34:7) and to the "litterarum monimenta" of the opening section of Cicero's *De inventione*, and is punctuated by the forms of prose rhythm (*cursus velox, tardus* and *planus*) that would later be formally recommended by the *artes dictandi.*[77] It shows that the study of rhetoric and classical authors, taught by Bernard at Saint-Martin and woven into the fabric of his *Cosmographia*, also served a more practical function in the preservation of his community's institutional memory.

It was only from the late eleventh century, first in Italy and then in the schools of the Loire valley, that the rules of the *ars dictaminis* came to be codified and transmitted through independent treatises that favoured utilitarian collections of model salutations, proverbs and letters over the more traditional and rhetorically theoretical approach, based on the study of ancient authors. The reasons for this shift in pedagogical emphasis lie in the growing use of written records by secular and ecclesiastical authorities in this period. This in turn was the result of a range of historical factors, as various as the stabilisation of feudal domains, the polemics of the Investiture controversy, the development of commerce and, especially in Italy, the bureaucratic demands of both the papacy and the emerging self-governing city states.[78] The *ars dictaminis* was then a potentially lucrative field of study. As Bernard of Meung, the author of an early French dictaminal treatise, wrote in 1187: "Truly those who pursue

76 *Gallia Christiana XIV*, ed. Hauréau, Instrumenta LX, 80AB.

77 *Cursus velox* ("aeternáliter rètinéri"), *tardus* ("quartámve progéniem", "custodíre memóriam"), and *planus* ("pósse deféndi").

78 M. Camargo, *Ars dictaminis, Ars dictandi* (Turnhout: Brepols, 1991), 31–32; R. Köhn, "Schulbildung und Trivium im lateinischen Hochmittelalter und ihr möglicher praktischer Nutzen", in *Schulen und Studium im sozialen Wandel des hohen und späten Mittelalters*, ed. J. Fried (Sigmaringen: Thorbecke, 1986), 203–84; J. O. Ward, "Rhetorical Theory and the Rise and Decline of *Dictamen* in the Middle Ages and Early Renaissance", *Rhetorica* 19 (2001): 175–223 (177–90).

the science of *dictamen* come to kings and are passed on by kings to the prelates of the Church, to be promoted to ecclesiastical honours on the strength of their service."[79]

With its tradition of expertise in the drafting of official documents, the school at Saint-Martin was well placed to meet this new demand for pragmatic instruction in letter-writing. Already in the 1040s and 1050s, after his tenure teaching the arts at Saint-Martin, Berengar of Tours found employment with Count Geoffrey Martel as an apologist for Angevin policy, writing for example to Pope Leo IX to justify Geoffrey's imprisonment of Bishop Gervase of Le Mans.[80] Later, studying at Tours in the 1140s or 1150s, Peter of Blois learned by heart the elegant letters of Hildebert of Lavardin, which Hildebert had begun to publish as a model collection in the 1120s.[81] After Tours, Peter went on to further study in Roman law at Bologna and theology at Paris, before leaving the schools to spend the rest of his career at courts in Sicily, France and eventually England, writing for various patrons letters such as those he sent to the pope to protest Richard the Lionheart's imprisonment in Germany.[82] From 1184, Peter, like Hildebert before him, began to assemble his letters into a model collection that proved extremely influential, surviving in some 250 manuscripts.[83]

Increasingly the forms and phrasing of such official correspondence came to be dictated by the new prescriptive manuals on the *ars dictaminis*. One such manual, the *Libellus de arte dictandi rhetorice*, which dates from the 1180s, was

79 "Verum enimvero qui dictandi secuntur scienciam ad reges veniunt et prelatis ecclesie traduntur a regibus, ad honores ecclesiasticos interventus sui potencia promovendi." The quotation from Bernard's *Flores dictaminum* is cited from Vienna, Österreichische Nationalbibliothek 521, f. 161v (under the title *Major Compilatio*) by C. Vulliez, "L'apprentissage de la rédaction des documents diplomatiques à travers l'ars dictaminis français (et spécialement Ligérien) du xiie siècle", in *Cancellaria e cultura*, ed. Gualdo, 77–96 (77).

80 Gibson, "Letters and Charters", 11–12.

81 *Ep.* 101 (to the archdeacon of Nantes), *PL* 207, 314A: "Profuit mihi, quod epistolas Hildeberti Cenomanensis episcopi styli elegantia, et suavi urbanitate praecipuas firmare, et corde tenus reddere adolescentulus compellebar." On the collection, see P. von Moos, *Hildebert von Lavardin, 1056–1133: Humanitas an der Schwelle des höfischen Zeitalters* (Stuttgart: Anton Hiersemann, 1965), 326–40.

82 R.W. Southern, "Peter of Blois", *Oxford Dictionary of National Biography*, 60 vols. (Oxford: Oxford University Press, 2004), VI: 242–47.

83 L. Wahlgren, *The Letter Collections of Peter of Blois: Studies in the Manuscript Tradition* (Gothenburg: Acta Universitatis Gothoburgensis, 1993), 17.

in all likelihood written by Peter of Blois.[84] The preface to this work, which explains Peter's reason for writing, is an important source for understanding the academic milieu at Tours. He writes:

> Licet magistri Bernardi de dictaminibus liber prudenter sit pertractus, delicatis tamen et minus discretis lectoribus perplexae prolixitatis dicitur arguendus. Turonenses eciam magistros dictandi scienciam in summulas redigentes nichil perfectum asserunt conscripsisse.[85]

> Although the book of Master Bernard on *dictamen* is intelligently composed, it is said that novices and undiscerning readers accuse it of being confused and too long. They also claim that the masters of Tours have not written fully on the science of *dictamen* in their summary treatises.

The most authoritative scholarship has argued that the Bernard referred to here is not Bernard Silvestris, but rather the Italian rhetorician Bernard of Bologna, whose *Summa dictaminum* was popular in the northern French schools, and a source for Peter's own *Libellus*.[86] On the other hand, there are various extant works that could be the Tours *summulae* to which Peter refers.[87] The most obvious candidate is an anonymous treatise entitled *Flores rhetorici*, which together with Bernard of Bologna's *summa* is one of the major sources for Peter's *Libellus*. Datable to *c.* 1160–1170, the *Flores rhetorici* could be called "incomplete" because it omits or abbreviates topics that were central to the

84 For discussion of the attribution, see M. Camargo, *Medieval Rhetorics of Prose Composition: Five English Artes Dictandi and their Tradition* (Binghamton, NY: Medieval and Renaissance Texts and Studies, 1995), 39–41. Further evidence in favour of the attribution is presented by Worstbrock *et al.*, *Repertorium*, 90–92, and in *Petri Blesensis Carmina*, ed. Wollin, 104–18.

85 *Libellus de arte dictandi rhetorice*, ed. Camargo, *Medieval Rhetorics*, 45.

86 M. Camargo, "The *Libellus de arte dictandi rhetorice* Attributed to Peter of Blois", *Speculum* 59 (1984): 16–41 (24–26), following C. H. Haskins, "An Italian Master Bernard", in *Essays in History Presented to Reginald Lane Poole*, ed. H. W. C. Davis (Oxford: Clarendon Press, 1927), 211–26.

87 Worstbrock *et al.*, *Repertorium*, propose other specific sources for Peter's treatise in the *Aurea gemma Gallica* (119–22), the *Forma dictandi* (134–35), and the *Summa dictaminis* attributed to Rudolph of Tours (96–98). See also F. J. Worstbrock, "Die Frühzeit der Ars dictandi in Frankreich", in *Pragmatische Schriftlichkeit im Mittelalter: Erscheinungsformen und Entwicklungsstufen*, ed. H. Keller, K. Grubmüller and N. Staubach (Munich: Fink, 1992), 131–56 (149–52).

new *ars dictaminis*—such as *salutatio* formulas—to focus instead on more literary concerns, especially metaphor, and to make extensive (and increasingly unfashionable) use of some of the classical authors brought back into circulation by the Carolingians. An example of the work's literary approach to *dictamen* can be seen in its relatively lengthy discussion of how to describe a *locus amoenus*. In particular, this passage contains a list of trees of a kind that is found in none of the extant treatises except for Matthew of Vendôme's *Ars versificatoria* and also in Bernard's *Cosmographia*, with both of which there are verbal parallels in this section of the *Flores*.[88]

This and other internal evidence strongly suggest that the *Flores* was composed in the orbit of the schools of Tours. However, the authorship of the treatise remains a matter of conjecture. Despite numerous parallels with the *Ars versificatoria*, which suggest that parts of that treatise were in circulation before the traditional date for the work of *c.* 1175, Matthew's authorship has rightly been questioned on the grounds of his differing citation technique and the notably divergent handling of the subject of metaphor in his *Ars*.[89] Another suggestion is that the *Flores* was in fact a youthful work by Peter of Blois, written while he was still at Tours, then updated around two decades later in his *Libellus*. However, whilst it is true that the *Flores* does contain certain echoes of poems attributed to Peter, these are not necessarily self-quotations. And although the treatise does indeed have many passages in common with the *Libellus*, this can be accounted for simply by the fact that it was one of the two major sources used by the author of the later work. More tentatively, Martin Camargo has written that "the most plausible candidate for the authorship of the *Flores* remains Bernard Silvestris. But the case for ascribing the treatise to him, based as it is on scant and indirect evidence, must be called weak at present."[90] The *Flores* does indeed quote Bernard's verse on a couple of occasions, but again this does not entail a single author as opposed to a school in which certain texts, ideas and examples were in ready circulation.[91]

88 M. Camargo, "A Twelfth-Century Treatise on 'Dictamen' and Metaphor", *Traditio* 47 (1992): 161–213 (205–8).

89 C. Wollin, "Beiträge zur Werkchronologie und Rezeption des Matthäus von Vendôme", *Sacris Erudiri* 45 (2006): 327–52.

90 Camargo, "Twelfth-Century Treatise", 169.

91 To the quotation identified by Camargo in his edition (*Flores rhetorici*, l. 391: "gigas abies, pi<g>mea mirica" = *Cos.* I.3.277), Wollin, "Werkchronologie", 331, n. 9, adds *Flores*, l. 141: "Denigrat meritum dantis mora" = *Math.* 763. Unnoticed by Wollin is the fact that this second quotation reappears in Peter's later *Compendium in Iob* (see above, n. 67); again though, this could equally well be a reminiscence of an exemplary proverb from Peter's schooldays at Tours as an instance of 'self-quotation' from the *Flores rhetorici*.

The attribution of the *Flores* to Bernard is further complicated by evidence that he wrote a rhetorical treatise that was much wider in scope. In his *Ars poetica* (1208–1216), the English rhetorician Gervase of Melkley remarks that Bernard had already written an extensive treatise on composition:

> Scripserunt autem hanc artem Matheus Vindocinensis plene, Gaufroi Vinesauf plenius, plenissime vero Bernardus Silvestris, in prosaico psitacus, in metrico philomena.[92]

> Matthew of Vendôme wrote fully about this art, Geoffrey of Vinsauf more fully still, but most fully of all Bernard Silvestris, a parrot in prose and a nightingale in verse.

As Douglas Kelly has remarked, it is strange that no surviving treatise on composition has been convincingly attributed to Bernard, given that the allegedly lesser works of Matthew and Geoffrey (the *Poetria nova*, 1208–1213) survive in a total of nineteen and two hundred manuscripts respectively.[93] Kelly has suggested that Bernard's supposedly lost treatise on the art of composition may in fact be the *Cosmographia* itself. To support this he points to another assessment of Bernard by the rhetorician Evrard of Béthune, who wrote in his *Laborintus* (before 1280) that: "One acquires these flowers [of rhetoric] through the art of *dictamen*: Bernard's major and minor summa will give you that art." Kelly suggests that this is not another reference to a lost *ars poetica*, but that the "maior summa minorque" in fact refers to the *Cosmographia*, with its two books on the macrocosm and the microcosm.[94] Furthermore, he proposes that

---

92 Gervase of Melkley, *Ars poetica*, ed. Gräbener, 1.9–11.

93 Kelly, *Arts*, 57–63. M. Brini Savorelli, "Il 'Dictamen' di Bernardo Silvestre", *Rivista critica di storia della filosofia* 20 (1965): 182–230, attributed to Bernard the treatsie *De competenti dictaminum et grata scientia* in Vienna, Österreichische Nationalbibliothek 246, but this has been shown to be a southern French treatsie composed between 1204 and 1214. See M. Klaes, "Die 'Summa' des Magister Bernardus: zu Überlieferung und Textgeschichte einer zentralen Ars dictandi des 12. Jahrhunderts", *Frühmittelalterliche Studien* 24 (1990): 198–234 (199–202, 232–33).

94 Evrard of Béthune, *Laborintus*, ll. 597–98, ed. E. Faral, *Les arts poétiques du xiie et du xiiie siècle: recherches et documents sur la technique littéraire du moyen âge* (Paris: Honoré Champion, 1962), 336–77 (358): "Hos via dictandi recipit flores: dabit illam | Bernardi maior summa minorque tibi." One problem with this is that Gervase later refers more explicitly to the *Cosmographia* (within a list of authorities that combine prose and verse) in somewhat different terms (ll. 683–84): "Tractatum titulos laudis Bernardus habentem | De gemina mundi partitione facit." Nonetheless, it is hard to see how any of the more

the mixed form of the *Cosmographia* was also what Gervase of Melkley had in mind when he compared Bernard favourably to his successors as a master of both poetry *and* prose. Given that Bernard is the most quoted author in Gervase's *Ars poetica*, this hypothesis is quite plausible. Not only do both the prose and verse of the *Cosmographia* illustrate many of the rhetorical figures discussed by Gervase, but the prose chapters provide an unusually consistent model of the forms of *cursus* that were recommended by authorities on the *ars dictaminis*.[95]

The twelfth-century abbey of Saint-Martin, then, thanks in part to its extensive wealth, had a long history of expertise in the composition of official documents. Due to their solemn nature, such documents were often ornamented with rhetorical figures that had traditionally been acquired through the study of classical theoreticians such as Cicero and the exemplary prose and verse of antiquity. With the expansion of bureaucracy in the eleventh and twelfth centuries, there was a new demand for more pragmatic letter-writing manuals that contained tailor-made salutations, proverbs and other formulas. At Tours there seems to have been an attempt, evident in the *Flores rhetorici*, to adapt the classical *auctores* and some of the more literary aspects of the rhetorical tradition (metaphor, *descriptio loci*) to these new educational requirements. It is in this sense that Martin Camargo has written of the "pedagogical conservatism" of the *Flores*, an adherence to a tradition of literary instruction that can be traced back to Alcuin and the Carolingian schools.[96] Both that work and Gervase of Melkley's *Ars poetica* suggest that the *Cosmographia* was

---

practical *artes dictandi* attributed to other Bernards (of Bologna, of Meung, *etc.*) could be said to provide exemplary illustration of the long list of figures of thought (*flores*) that Gervase had just enumerated (ll. 523–96).

95 T. Janson, *Prose Rhythm in Medieval Latin: From the Ninth to the Thirteenth Century* (Stockholm: Almqvist & Wiksell, 1975), 75, 113, observes that over 96% of Bernard's prose sentences end with a rhythmical *clausula*. Bernard also used rhythm to emphasise syntactic structure within his sentences. In *Cos.* I.2, for example, there are fifty-five instances of *cursus velox* within (rather than at the ends of) sentences, as these sentences are punctuated by Dronke. All but around eight of these mark some kind of syntactic break. Internal rhythm has been little studied, perhaps because it is less susceptible to the kind of statistical analysis that has been applied to rhythms at the ends of sentences. For some preliminary remarks on the relationship between internal rhythm and medieval manuscript punctuation, see G. Orlandi, "Metrical and Rhythmical Clausulae in Medieval Latin Prose: Some Aspects and Problems", in his *Scritti di filologia mediolatina*, ed. P. Chiesa, A. M. Fagnoni, R. E. Guglielmetti, and G. P. Maggioni (Florence: SISMEL, 2008), 513–32 (523, 531–32).

96 Camargo, "Twelfth-Century Treatise", 170.

involved in this programme. It will be important to keep in mind this idea of the *Cosmographia* as a work that bridged a gap between older and newer forms of learning when we come, in the following two chapters, to explore the ways in which Bernard's work drew on the classical tradition to address some major intellectual concerns of the twelfth-century schools.

Poetry and Imitation

Before turning to the *Cosmographia*, we should conclude by examining the handful of twelfth-century poems that appear in the final folios of the aforementioned Saint-Martin schoolbook: Tours, Bibliothèque municipale 300. These poems illuminate some of the ways in which ancient sources were absorbed, imitated and adapted in the school milieu in which Bernard worked. As we have seen, training in prose composition for charter and letter writing had an obvious vocational dimension, for it could lead to employment in the burgeoning administrative structures of twelfth-century Europe. The same could be said for many of the other advanced subjects studied in the schools. Logic and law, for example, would have complemented the *ars dictaminis* as disciplines relevant to various administrative positions.[97] Medicine had obvious practical applications, and physicians trained in the more theoretical medicine of the Arabo-Latin translations were increasingly in demand at European courts.[98] Theology too was often a path to higher ecclesiastical positions in the episcopacy or papal curia.[99] By contrast, training in verse composition might not appear to have such immediate uses. How then is the study of versification in the schools to be understood?

The first thing to note is that the boundaries between verse and prose were much more porous than they are today. On the one hand, prose authors who employed the rhythmical cadences of *cursus* required a knowledge of vowel quantities that was also necessary for writing verse.[100] On the other hand,

97 A. Murray, *Reason and Society in the Middle Ages* (Oxford: Clarendon Press, 1978), 213–33.

98 E.J. Kealey, *Medieval Medicus: A Social History of Anglo-Norman Medicine* (Baltimore, MD: Johns Hopkins University Press, 1981).

99 Southern, *Scholastic Humanism*, I: 163–85.

100 In a period before dictionaries recorded vowel quantities, such knowledge was transmitted mainly through systems for learning the quantities of first, middle and final syllables. See J. Leonhardt, *Dimensio syllabarum: Studien zur lateinischen Prosodie- und Verslehre von der Spätantike bis zur frühen Renaissance* (Göttingen: Vandenhoeck und Ruprecht, 1989).

verse was not considered an optional or advanced component of grammatical education; on the contrary, many of the standard works that introduced pupils to the Latin language were poetic compositions—texts such as the Psalter, the *Aeneid*, and the *Disticha Catonis*.[101] One could even make the case that scholars generally left behind youthful school exercises in verse composition to concentrate in their maturity on more substantial prose works of theology or polemic.[102] Furthermore, there was not yet the Romantic notion of poetry as a sphere of discourse especially suited to self-expression, and separated from the prosaic communication of other more impersonal information. It is certainly true that the twelfth century saw a remarkable flowering of lyric verse, that expressed more personal concerns, and that clerics from schools like Saint-Martin could obtain the patronage of secular and ecclesiastical lords seeking to cultivate the new courtly ideals of the period.[103] Nonetheless, verse continued to be used, as it had since antiquity, for a wide range of didactic and technical literature, including theology, hagiography, history, and rhetoric.[104]

On one level, the use of verse in such contexts was a practical convenience. Students were often called on to memorise sections of their textbooks, and the mnemonic qualities of verse could make this an easier task.[105] More fundamentally, the versification of broadly didactic material should be understood in terms of the Ciceronian notion of the interdependence of *sapientia* and *eloquentia*, the idea that what is true should be well expressed.[106] In some

---

101 P. Stotz, "Dichten als Schulfach—Aspekte mittelalterlicher Schuldichtung", *Mittellateinisches Jahrbuch* 16 (1981): 1–16 (7).

102 J. Van Engen, "Letters, Schools, and Written Culture in the Eleventh and Twelfth Centuries", in *Dialektik und Rhetorik im früheren und hohen Mittelalter: Rezeption, Überlieferung und gesellschaftliche Wirkung antiker Gelehrsamkeit vornehmlich im 9. und 12. Jahrhundert*, ed. J. Fried (Munich: Oldenbourg, 1997), 97–132 (101–2).

103 C. J. Mews, *The Lost Love Letters of Heloise and Abelard: Perceptions of Dialogue in Twelfth-Century France* (New York: St. Martin's, 1999), 87–114; G. A. Bond, *The Loving Subject: Desire, Eloquence, and Power in Romanesque France* (Philadelphia: University of Pennsylvania Press, 1995), 66–69, 136.

104 T. Haye, *Das lateinische Lehrgedicht im Mittelalter: Analyse einer Gattung* (Leiden: Brill, 1997).

105 M. C. Woods, *Classroom Commentaries: Teaching the "Poetria nova" across Medieval and Renaissance Europe* (Columbus: Ohio State University Press, 2010), 21–22, suggests that this may account for the popularity over their prose rivals of versified manuals such as Geoffrey of Vinsauf's *Poetria nova*, Evrard of Béthune's *Laborintus*, and Alexander de Villa Dei's *Doctrinale*.

106 Cicero, *De inventione* I.1: "Ac me quidem diu cogitantem ratio ipsa in hanc potissimum sententiam ducit, ut existimem sapientiam sine eloquentia parum prodesse civitatibus, eloquentiam vero sine sapientia nimium obesse plerumque, prodesse numquam." See

cases there was ambivalence about using the metrical forms of pagan antiquity to adorn Christian doctrine. Thus Andrew Sunesen, Archbishop of Lund (d. 1228), remarked in the preface to his hexameter poem on the first six days of creation that although metre is associated with the *vana poetarum figmenta*, it is nonetheless integral to good Latin instruction.[107] The seductive verse of the ancients may lead the student into the errors of pagan fable, but scriptural material, if ornamented by metre, can cure this delusion and lead him back to salvation.[108] In other cases, the elevated role of ancient *eloquentia* was accepted unreservedly. John of Salisbury's *Metalogicon* (*c.* 1159), for example, famously advocates the study of Vergil and Lucan as *auctores* who had adorned the basic narratives of their works with "such a great wealth of learning . . . that the finished masterpiece (*opus consummatum*) would seem to image all the arts".[109] The *Cosmographia* itself seems to have aspired to the status of such an *opus consummatum*, through the topical deployment of parallels between the processes of cosmic and poetic creation, of a kind found also in Plato, Ovid and other ancients.[110] Thus, just as John of Salisbury wrote of the "rudem materiam" of the basic narrative of events that the *auctores* transformed through the imposition of rhetorical forms into an *opus consummatum*, so Bernard wrote of the "rudes . . . materias" that the Mind of God shaped through the imposition

---

further J. O. Ward, *Ciceronian Rhetoric in Treatise, Scholion and Commentary* (Turnhout: Brepols, 1995), 274–77.

107 *Andreae Sunonis Filii Hexaemeron*, ed. S. Ebbesen and L. B. Mortensen, 2 vols. (Copenhagen: Gad, 1995–1998), I: 76, ll. 32–46.

108 *Hexaemeron*, ed. Ebbesen and Mortensen, I: 77, ll. 75–76: "Tot morbis sacra posset scriptura mederi, | si depicta metri vernantis floribus esset"; and I: 78, ll. 104–5: "Tam dulcem pareret fructum coniunctio metri: | alliceret iuvenes ad dogma salutis . . ."

109 John of Salisbury, *Metalogicon* I.24, ed. Hall and Keats-Rohan, 52.27–33: "Illi enim per diacrisim quam nos illustrationem sive picturationem possumus appellare, cum rudem materiam historiae aut argumenti aut fabulae aliamve quamlibet suscepissent, eam tanta disciplinarum copia, et tanta compositionis et condimenti gratia excolebant, ut opus consummatum omnium artium quodam modo videretur imago." These ideals were based on Book I of Quintilian's *Institutio oratoria*: see P. Godman, "*Opus consummatum, omnium artium . . . imago*: From Bernard of Chartres to John of Hauvilla", *Zeitschrift für deutsches Altertum und deutsche Literatur* 124 (1995): 26–71 (30–31).

110 On the tradition of the 'poème-univers', see P. Hadot, *Le voile d'Isis: essai sur l'histoire de l'idée de Nature* (Paris: Gallimard, 2004), 214–18; and for specific moments in its ancient development: L. Brisson, "Le discours comme univers et l'univers comme discours: Platon et ses interprètes néo-platoniciens", in *Le texte et sa représentation* (Paris: École Normale Supérieure, 1987), 121–27; G. Lieberg, *Poeta Creator: Studien zu einer Figur der antiken Dichtung* (Amsterdam: Gieben, 1982); and J. A. Coulter, *The Literary Microcosm: Theories of Interpretation of the Later Neoplatonists* (Leiden: Brill, 1976), 95–126.

of *numeri*.[111] These *numeri* are the arithmetical bonds of the elements, but a *numerus* could also be a poetic measure, and there is an analogy running through the *Cosmographia* between the author's rhetorical ornamentation of his subject matter and the divine mind's fashioning of primordial matter into a "consummatum ... conpositum" (*Cos.* I.2.11.5). The well-ordered treatise—one bound by the harmonies of metre—is the one that will give the best image of the well-ordered cosmos that it aims to represent.

A number of the pedagogic foundations on which the lofty edifice of the *Cosmographia* was built can be seen in the final folios of Tours, Bibliothèque municipale 300. These contain four poems that rework ancient sources in ways that set the scene for the *Cosmographia*'s handling of philosophical and literary traditions. One of these poems is by Peter of Saintes, a tutor to the English king Henry II. Known from its opening words as "Viribus, arte, minis", the poem is a 122-line composition that narrates the origins of the Trojan war, the siege and destruction of Troy, and Aeneas's foundation of Rome.[112] As a paraphrase of Vergil's *Aeneid*, Peter's poem is an example of a form of cultural appropriation that Rémi Brague has called "digestion", where the paraphrasing of an authoritative text is an indicator of the depth of its absorption by the host culture.[113] Yet even Peter's relatively simple poem shows how the authority of ancient models did not have to exert a constraining influence on later writers. Take this description of Helen's beauty and its effects:

> Crines, colla, gene, cunctaque compta bene.
> Quam facit audaces amor in sua damna procaces.[114]

---

111 *Cos.* I.2.7.6–8: "Rudes ut erant, indisciplinatas reluctantesque materias exequavit potentiis, coniunxit medietatibus, numeris illigavit."

112 The text in *PL* 171, 1451B–1453A, prints the poem as an uninterrupted continuation of a Troy poem by Simon Chevre d'Or. The poem is printed independently by E. du Méril, *Poésies populaires latines antérieures au douzième siècle* (Paris: Brockhaus et Avenarius, 1843), 400–5. Du Méril notes only some of the variants in the many manuscripts of the poem.

113 R. Brague, "Inclusion and Digestion: Two Models of Cultural Appropriation in Response to a Question of Hans-Georg Gadamer (Tübingen, September 3, 1996)", in his *The Legend of the Middle Ages: Philosophical Explorations of Medieval Christianity, Judaism, and Islam*, trans. L. G. Cochrane (Chicago: University of Chicago Press, 2009), 145–58. Brague's examples of digestion are the paraphrases of Aristotle by Avicenna and Albert the Great.

114 Here I follow the text in *PL* 171, 1451C—substantially the same in du Méril (402)—rather than the inferior version of the second line in Tours, Bibliothèque municipale 300, f. 74v: "Quos facit audaces facit in sua damna procaces."

Hair, neck and cheeks—all was coiffed and neat.
Oh, how love makes the bold thirst for their doom to unfold!

Apart from the framing of this material in the medieval form of elegiacs with internal rhyme, these lines demonstrate two other characteristics of medieval versification, namely the brief *descriptio puellae* (*cf.* Matthew of Vendôme, *Ars versificatoria* I.56) and the moralising *sententia* of the second line.[115] Beyond such rhetorical reworking, the poem also departs for the narrative of its source in one striking point of detail: although Peter concludes with certain details of Aeneas's post-Trojan career—the visit to the Cumaean Sibyl, the war with Turnus, and the foundation of Rome—he omits all reference to Dido, a not insubstantial character in Vergil's poem! Christopher Baswell has convincingly suggested that, as a tutor to Henry II, himself the ruler of a European empire, Peter may have excised Dido from the story in order to create a more suitable role model for the young king, namely an Aeneas "of uninterrupted imperial intention and marital purity."[116]

Peter of Saintes had no known connections with Tours, but another poem in the final folios of the manuscript was written by the future archbishop of the city, Hildebert of Lavardin, while he was still bishop of Le Mans, in 1099 or 1100. Hildebert wrote this poem, known by the titles *De casu huius mundi* and *De exilio suo*, whilst in England, where he had been forced to travel with King William II after members of his cathedral clergy accused Hildebert of conspiring with Count Helias of Le Mans to expel the garrison that William, Helias's Norman overlord, had imposed on the city.[117] The ninety-line poem may be divided into three equal parts: a description of how Fortune once favoured the poet (1–30); the reversal of his fortunes and the storm that struck the poet's ship as he sailed to exile in England (31–60); and concluding reflections on the unity of divine purpose that lies behind the vicissitudes of human life (61–90). Like Peter of Saintes, Hildebert adapted ancient material for a contemporary purpose. But where Peter rather crudely refashioned his Vergilian source for a didactic end, Hildebert made more artistic use of ancient models to express an emotional, and ultimately philosophical response to his predicament.

115 The text in the Tours manuscript emphasises the morally edifying purpose of the poem by moving line 13 in the printed editions to the head of the text: "Ne quis amet temere docet obruta Troia cavere."

116 C. Baswell, *Virgil in Medieval England: Figuring the "Aeneid" from the Twelfth Century to Chaucer* (Cambridge: Cambridge University Press, 1995), 178.

117 Von Moos, *Hildebert von Lavardin*, 8–9.

In the first two thirds of his poem Hildebert is much indebted to Ovid's exilic poetry, for both his characterisation of Fortune's fickleness and the dramatic description of the storm at sea. There are also reminiscences of the storm scene in *Aeneid* I, for example of the North Wind raising waves to the heavens (*Aen.* I.103: "fluctusque ad sidera tollit"):

> Ecce rapax turbo, tollens ad sidera fluctus,
> Impulit ad litus iam sine puppe ratem.[118]

> Lo! how the raging storm, lifting waves to the stars,
> Drove to the shore the now rudderless raft.

In the final third of his poem, Hildebert is indebted to another school author, Boethius, for his conclusion that what appears to human perception as the random workings of Fortune are in fact varied manifestations of the divine will:

> Si fas est credi te quicquam posse vel esse,
> O Fortuna, quod es, quod potes ipse dedit.[119]

> Fortune, if it's right that you are whatever is or can be,
> It is He who has given you power and being.

One way of reading Hildebert is as a pure 'school poet', drawing on ancient philosophy and verse merely to achieve an effect of literary polish.[120] Yet Hildebert's contemporaries had a rather higher estimate of his abilities. Orderic Vitalis, for example, wrote that Hildebert "was a poet without peer in our time, and composed many songs that equalled or surpassed the poems of the ancients, which learned men eagerly aspire to explain with subtlety, and are more anxious to acquire for themselves than gold or topaz."[121] It is true that depictions

118 Tours, Bibliothèque municipale 300, f. 75r. Both this and the following quotation agree with the text in Hildebert of Lavardin, *Carmina minora* 22.57–58, ed. A. B. Scott (Leipzig: Teubner, 1969), 13.

119 *Carm.* 22. 85–86; *cf.* Boethius, *De consolatione Philosophiae* III, pr. 12.7, ed. Bieler, 60.20–21: "... nisi unus esset qui has mutationum uarietates manens ipse disponeret."

120 Godman, "*Opus consummatum*", writes of "the empty virtuosity of Loire mannerism" (54), and of Hildebert, Baudri of Bourgueil and Marbod of Rennes as "an over-rated circle of poetasters" (38).

121 *Historia Ecclesiastica* X.iv.42, in *The Ecclesiastical History of Orderic Vitalis*, ed. and trans. M. Chibnall, 6 vols. (Oxford: Clarendon Press, 1969–1980), V: 236: "Temporibus nostris

of storms at sea and shipwrecks were part of a literary tradition stretching back to Ovid, Vergil, and beyond. But this particular poem is no mere school exercise. Hildebert really did have to cross the Channel in exile, and his account of the experience is far from the *cento* ('patchwork') of Vergilian hexameters that make up the storm scene in *Apollonius of Tyre*.[122] Rather Hildebert turned instinctively to ancient authors to find a language rich enough in imagery and ideas to express a complex response to his predicament. The terrifying drama of the storm gives way to verses that do not simply recycle Boethian platitudes, but rather model a tone of aesthetic and moral calm.[123] The language may be borrowed, but the voice is Hildebert's own.

The examples of Peter and Hildebert help us see how, among Bernard's contemporaries, ancient sources were reworked to serve the needs of the present—in the first case as a form of advice literature, in the second as a means to articulate a compelling ethical standpoint. A third form of engagement with antiquity is on display in the two remaining poems from the end of the Tours manuscript. The *Mathematicus*, which we have seen was most probably written by Bernard, tells the story of a man foreseen by an astrologer to become emperor of Rome, and to kill his own father. The other poem, *De gemellis*, recounts the circumstances of a court case brought by a woman against her husband, who had taken the advice of doctors that the only way to treat their ailing twins was to dissect one in order to discover a cure for the other. Both poems are based on the late antique *Declamationes maiores* falsely attributed to Quintilian, the *Mathematicus* on the fourth declamation, and *De gemellis* on the eighth. Each of the pseudo-Quintilianic declamations begins with the summary of a situation in which a morally difficult decision must be made, sometimes with reference to particular Roman laws on the subject. The declamation itself then consists of a speech advocating a particular course of action—for example, in the fourth declamation that the senate should allow the emperor to suicide, or in the eighth that the father was guilty of mistreating the dead twin.[124]

---

incomporabilis uersificator floruit, et multa carmina priscis poematibus æqualia uel eminentia condidit; quae feruidus calor philosophorum subtiliter rimari appetit, ac super aurum et topazion sibi consciscere diligenter satagit."

122 G. A. A. Kortekaas, *Commentary on the Historia Apollonii Regis Tyri* (Leiden: Brill, 2007), 141.

123 Here I follow von Moos, *Hildebert von Lavardin*, 102, who refuses to see the poem as "[eine] elegante Schulübung".

124 For the texts, see *Declamationes XIX maiores Quintiliano falso ascriptae*, ed. L. Håkanson (Stuttgart: Teubner, 1982), 60–84 (IV), 151–74 (VIII).

Declamation played an important role in ancient Roman education, initiating students into the oratorical culture that was a prominent feature of civic life; and an emphasis on declamatory training appears in those Roman manuals that shaped the study of rhetoric in the Middle Ages: Cicero's *De inventione*, and the pseudo-Ciceronian *Rhetorica ad Herennium*.[125] Yet exactly what roles Roman declamatory literature played in medieval education is unclear. There is scattered evidence from the tenth and eleventh centuries that the *Declamationes maiores* were used for training in public oratory, for example at the school of Reims under Gerbert of Aurillac.[126] However, it was in twelfth-century France that the works of pseudo-Quintilian attained the height of their medieval popularity. Of the twenty-one copies that survive from between the ninth and twelfth centuries, up to fourteen may have been produced in twelfth-century France—and to scholars of this period the name of Quintilian was synonymous with *causae* (lawsuits).[127] An intriguing product of this interest was a minor genre of twelfth-century poems based on the ancient declamations. Thus far seven such poems have been identified. Four of these—*Mathematicus, De gemellis, De paupere ingrato*, and *De nato Ethiope*—have already been mentioned in connection with the schools of Tours.[128] The remaining three poems have been attributed to Peter Riga, a canon at Reims in the final third of the twelfth century. One, *Sponsus adversus sponsam*, based on Seneca, *Controversiae* II.7, also appears in two twelfth-century *Cosmographia*

---

125 A. Corbeill, "Rhetorical Education and Social Reproduction in the Late Republic and Early Empire", and W. M. Bloomer, "Roman Declamation According to the Elder Seneca and Quintilian", in *A Companion to Roman Rhetoric*, ed. W. Dominik and J. Hall (Oxford: Blackwell, 2007), 67–82, 297–306.

126 T. Haye, *Lateinische Oralität: gelehrte Sprache in der mündlichen Kommunikation des hohen und späten Mittelalters* (Berlin: Walter de Gruyter, 2005), 11, 62–65, 140–42, 171–72, 189–90, 210–12, 231; Ward, *Ciceronian Rhetoric*, 121–22.

127 B. Munk Olsen, *L'étude des auteurs classiques latins aux xie et xiie siècles*, 3 vols. (Paris: CNRS, 1982–1989), II: 289–305; Alan of Lille, *Anticlaudianus* III.233–35, ed. R. Bossuat (Paris: Vrin, 1955), 95: "Quintilianus adest quadam sub imagine ueri | Causarum uelans umbras, litesque nouellas | Fingit et in litem cogit sine lite uenire"; Stephen of Rouen, *Draco Normannicus*, Appendix 1, ll. 25–28, in *Chronicles of the Reigns of Stephen, Henry II, and Richard I*, ed. R. Howlett, 4 vols. (London: Longman, 1884–1889), II: 767: "Hic Paris in facie, statura Nestor, Ulixes | Consilio, belli viribus Hector erat: | Croesus divitiis, in causis Quintilianus, | Eloquio Cicero, versibus ipse Maro." For Stephen of Rouen's extensive engagement with Quintilian, see J. O. Ward, "Quintilian and the Rhetorical Revolution of the Middle Ages", *Rhetorica* 13 (1995): 231–84 (253–62).

128 See 28, above.

manuscripts.[129] The others, known as *Apes pauperis* and *Causa inter pauperem et divitem de apibus*, are both based on pseudo-Quintilian's thirteenth declamation, in which a pauper seeks redress for the poisoning of his bees by a rich neighbour.[130] *Apes pauperis* appears in an important florilegium of classical and medieval Latin verse from the cathedral of Saint-Maurice, Tours, compiled perhaps between 1175 and 1180.[131] It would seem, then, that Tours was a significant centre for the writing and study of this declamation-inspired verse.

The place of such poetry in a school context may be surmised from John of Salisbury's description of the teaching methods of Bernard of Chartres, a description that he must have come by second-hand, perhaps through William of Conches, who was Bernard's student and John's teacher. According to John, Bernard of Chartres' Latin instruction rested on three main activities: the careful grammatical and rhetorical exposition of the *auctores*, the memorisation and recitation of passages from the day's reading, and "introductory exercises" (*praeexercitamina*) in which his pupils would compose poetry and prose in imitation of the authors they had been studying.[132] The poems based on prose declamations could be just such school compositions. They treat what Matthew of Vendôme called *materia illibata*, material previously "untouched" by the poets; and they perhaps served a purpose similar to that of the metrical

---

129 *Incipit*: "Rarius in terris nihil est quam femina recti"; *PL* 171, 1453B–1455C (where the poem is falsely attributed to Hildebert of Lavardin). The *Cosmographia* manuscripts are Berlin, Staatsbibliothek zu Berlin-Preussischer Kulturbesitz, Phillips 1694 and Reims, Bibliothèque municipale 1275. The verse anthology in the Reims manuscript is apparently based on that in the Berlin one: see C. J. McDonough, "Hugh Metel and the Floridus Aspectus of Peter Riga (Staatsbibliothek zu Berlin Preussischer Kulturbesitz Phillips 1694)", *Mediaeval Studies* 67 (2005): 27–74.

130 *Apes pauperis* begins: "Vestra peritia dum regit omnia, sidera tangit." The text in *PL* 171, 1400B-1402B (again spuriously attributed to Hildebert) has been reprinted with source annotations by G. Krapinger, "Die Bienen des armen Mannes in Antike und Mittelalter", in *Theatron. Rhetorische Kultur in Spätantike und Mittelalter / Rhetorical Culture in Late Antiquity and the Middle Ages* (Berlin: Walter de Gruyter, 2007), 189–201. The *Causa* begins "Vos, sancti patres, moveat mea quaestio, vestrum", and has been edited by A. Boutemy, "Recherches sur le *Floridus Aspectus* III", *Latomus* 8 (1949): 283–301 (292–97).

131 Tours, Bibliothèque municipale 890. See A. Wilmart, "Le florilège de Saint-Gatien: contribution à l'étude des poèmes d'Hildebert et de Marbode", *Revue Bénédictine* 48 (1936): 3–40, 147–81, 235–58.

132 John of Salisbury, *Metalogicon* I.24, ed. Hall and Keats-Rohan, 54.109–11: "Et quia in toto praeexercitamine erudiendorum nihil utilius est quam ei quod fieri ex arte oportet assuescere, prosas et poemata cotidie scriptitabant..." The term derives from Priscian's *Praeexercitamina*, ed. H. Keil, *Grammatici Latini*, 7 vols. (Leipzig: Teubner, 1855–1880), III: 430–40.

saint's life, which was sometimes composed by advanced students, using prose *vitae* for sources, as a kind of final dissertation at the end of their studies.[133]

In the *Institutio oratoria*, imperfectly transmitted in the Middle Ages but like the *Declamationes* popular in the twelfth century, Quintilian remarked that while some masters give their students only brief hints ("brevia ... vestigia", II.6.5) about how to develop the theme of the declamation, others "not content with giving directions for the arrangement of the subjects that they were setting for their pupils to declaim, worked them out at some length themselves."[134] It is possible, then, that some or all of these declamatory poems—and most probably the long *Mathematicus*—were magisterial demonstrations of how the exercise should be approached. Furthermore, the textual history of *De gemellis* indicates that these poems were not always read as finished products. Six of the eleven *De gemellis* manuscripts show verbal and structural variations that cannot be accounted for simply as mechanical errors introduced by scribal transmission. Rather, Robert Edwards has argued that later copyists used the poem as a kind of rhetorical workshop to practise techniques of abbreviation, amplification, and the alteration of meaning through verbal substitution.[135] Copies of the poem that are closer to the original are to be found in the remaining five manuscripts, of which three also include Bernard's *Cosmographia*, and one is the Tours manuscript under discussion.[136] This suggests that the Tours manuscript contains a version of the eighth declamation—by Bernard or a member of his school—that was sent into circulation with the *Cosmographia*, and later adapted by other scholars and students.

Like the poems by Hildebert and Peter of Saintes in the final folios of the Tours manuscript, the *Mathematicus* and *De gemellis* adapt ancient sources to particular ends. The declamations, however, were not used for a didactic purpose or to mediate personal experience, but rather as a way of exploring philosophical problems within the wider milieu of the twelfth-century schools.

---

133 Matthew of Vendôme, *Ars versificatoria* IV.3, ed. Munari, III: 194: "Amplius, materia de qua aliquis agere proponet aut erit illibata aut ab aliquo poeta primitus executa." See also F. Dolbeau, "Un domaine négligé de la littérature médiolatine: les textes hagiographiques en vers", *Cahiers de civilisation médiévale* 178 (2002): 129–39.

134 *Institutio oratoria* II.6.1, ed. and trans. H. E. Butler, 4 vols. (Cambridge, MA: Harvard University Press, 1966–1969), I: 258–59: "quidam materias, quas discipulis ad dicendum dabant, non contenti divisione dirigere latius dicendo prosequebantur" (translation adapted). Munk Olsen, *Étude des auteurs classiques*, II: 289–305, lists ten manuscripts of the *Institutio oratoria* of possible French origin from the ninth to the twelfth centuries.

135 Edwards, "Poetic Invention", 202–5.

136 The *Cosmographia* manuscripts are: Bodl. Rawlinson G 109, BN lat. 6415, and Vatican City, Biblioteca Apostolica Vaticana Reg. lat. 370.

De gemellis, which describes the controversial diagnosis and treatment of sick twins, is a poem concerned with what would now be called medical ethics, and one of a kind that was thrown into sharper relief by the Arabo-Latin medical translations that were starting to be studied in the northern French schools. Likewise the *Mathematicus*, a tale of an astrologer's prophecy and its consequences, engages with questions of astral determinism and free will that had become more pressing with the reception of new astrological translations in the same period. These are poems that take us beyond the primarily rhetorical and literary concerns of the schools of Tours that we have explored in this chapter. To read them properly, we must first understand the scientific programme of the *Cosmographia*—and to do that we must look beyond Tours to the wider intellectual landscape in which Bernard was working.

CHAPTER 2

The Science of the *Cosmographia*

The *Cosmographia* was in many ways a product of the traditions of literary *imitatio* practised at Tours and discussed in the previous chapter. It cannot, however, be fully understood in those terms alone. For the scientific themes of the work also connect it to developments in natural philosophy that were taking place in other European centres of learning in this period. The twelfth century saw the growth of a philosophical interest in nature that differed both from the mere accumulation of scientific lore in early medieval encyclopaedias and again from the moral allegorisation of the natural world found in patristic theology.[1] Inspired by the teachings of Plato's *Timaeus*, accessible primarily through Calcidius's partial translation and commentary, scholars were coming to see the causes of natural phenomena as ontologically separate from the divine will. The *opera naturae* were distinct from the *opera Dei*, and the natural world was subject to what Plato called a "law-based cause and reason", which could be grasped by human understanding.[2] The medieval development of these ideas can be seen most vividly in a number of texts from the northern French schools in the years immediately preceding the writing of the *Cosmographia*. On the one hand, around 1140, there appeared perhaps the first systematic Latin commentary on the *Timaeus*, by William of Conches; on

1 See the pioneering studies from the 1950s by Chenu, "Découverte de la nature", and Gregory, *Anima mundi*. To these should be added: M.-D. Chenu, "Nature and Man: The Renaissance of the Twelfth Century", in his *Nature, Man and Society in the Twelfth Century: New Theological Perspectives in the Latin West*, ed. and trans. J. Taylor and L. K. Little (Chicago: Chicago University Press, 1968), 1–48; T. Gregory, "La nouvelle idée de nature et de savoir scientifique au xiie siècle", in *The Cultural Context of Medieval Learning*, ed. J. E. Murdoch and E. D. Sylla (Dodrecht: Reidel, 1975), 193–218; A. Speer, *Die entdeckte Natur: Untersuchungen zu Begründungsversuchen einer 'scientia naturalis' im 12. Jahrhundert* (Leiden: Brill, 1995); and P. E. Dutton, "The Materialization of Nature and of Quaternary Man in the Early Twelfth Century", in *Man and Nature in the Middle Ages*, ed. S. J. Ridyard and R. G. Benson (Sewanee, TN: University of the South Press, 1995), 137–56. Some of the claims originally made by Chenu have been called into question by S. Ritchey, "Rethinking the Twelfth-Century Discovery of Nature", *Journal of Medieval and Early Modern Studies* 39 (2009): 225–55.

2 Calcidius, *Commentarius* XXIII, in *Timaeus a Calcidio translatus commentarioque instructus*, ed. J. H. Waszink (London: Warburg Institute, 1962), 73.10–12: "Omnia enim quae sunt vel dei opera sunt vel naturae vel naturam imitantis hominis artifices"; and *Timaeus* 28A, ed. Waszink, 20.21–22: "nihil enim fit, cuius ortum non legitima causa et ratio praecedat."

 | DOI 10.1163/9789004268357_004

the other, the cosmological principles discussed by Plato and Calcidius were applied to the Christian study of creation in innovative commentaries on the first six days of Genesis by Peter Abelard and Thierry of Chartres, both from the 1130s.

In addition to these Platonic sources, there was also the stimulus of more recently available Latin translations of Arabic science from the peripheries of Christendom, works that transmitted and developed ancient philosophical ideas previously unknown in the Latin West.[3] The first wave of medical texts from southern Italy, in translations produced primarily by Constantine the African (*c.* 1020–1087), introduced a new level of theoretical sophistication to European medicine.[4] The cathedral school at Chartres appears to have been an important centre for the reception of such works.[5] One of the school's most famous students, William of Conches, was the contemporary of Bernard's to make the greatest philosophical use of such material, even employing medical concepts to explain cosmological phenomena in his *Dragmaticon philosophiae* (1144–1149).[6] At the same time, new translations of Arabic astrology, most prominently Abu Ma'shar's *Introductorium maius* in Latin versions by John of Seville (1130) and Hermann of Carinthia (1140), were starting to have an impact on the natural philosophy of the French schools.[7] The *Introductorium maius* presented the first theoretical explanation of celestial influence in Latin, and the independent treatise by one of its translators, Hermann of Carinthia's

3 See above, Introduction, n. 38.

4 D. Jacquart and F. Micheau, *La médecine arabe et l'Occident médiéval* (Paris: Maisonneuve et Larose, 1990), 87–129; N. G. Siraisi, *Medieval and Early Renaissance Medicine: An Introduction to Knowledge and Practice* (Chicago: University of Chicago Press, 1990), 1–16.

5 C. Burnett, "The Contents and Affiliation of the Scientific Manuscripts Written at, or Brought to, Chartres in the Time of John of Salisbury", in *The World of John of Salisbury*, ed. M. Wilks (Oxford: Blackwell, 1988), 127–60 (139–40); E. Wickersheimer, *Les manuscrits latins de médecine du haut moyen âge dans les bibliothèques de France* (Paris: CNRS, 1966), 17–30.

6 D. Jacquart, "Les emprunts de Guillaume de Conches aux théories médicales", in *Guillaume de Conches: philosophie et science au xiie siècle*, ed. B. Obrist and I. Caiazzo (Florence: SISMEL, 2011), 79–110; I. Ronca, "The Influence of the *Pantegni* on William of Conches's *Dragmaticon*", in *Constantine the African and ʿAlī ibn al-ʿAbbās al-Magūsī: The "Pantegni" and Related Texts*, ed. C. Burnett and D. Jacquart (Leiden: Brill, 1994), 266–85; D. Elford, "William of Conches", in *History of Twelfth-Century Western Philosophy*, ed. Dronke, 308–27.

7 Lemay, *Abu Ma'shar*. For debate over some of the details of Lemay's thesis, see P. Dronke, "New Approaches to the School of Chartres", *Anuario de estudios medievales* 6 (1969): 117–40 (128–31); Elford, "William of Conches", 312; and (with specific reference to the *Cosmographia*) below, 69–75.

De essentiis (1143), offered a bold and novel cosmology framed largely in terms of celestial science.

This astrological and medical learning generated an increasingly Aristotelian conception of *physica*, which could not easily be accommodated to the traditional, 'Platonic' equation of that science with the purely numerical quadrivium of the seven liberal arts.[8] Instead, Boethius's previously neglected Aristotelian definition of *physica* as the study of "the forms of bodies together with their constituent matter" began to attract increasing attention.[9] This development may be observed in a striking change of emphasis in William of Conches's *Timaeus* commentary from an earlier commentary, attributed by some to William's master, Bernard of Chartres (*fl.* 1100–1120), and described by others as an accretion of various eleventh-century glosses.[10] The prologue to the earlier commentary remarks that the *Timaeus* concerns *physica*, "since it speaks of plane figures and solid bodies, the embodiment of the world soul and other souls, their perpetual motion, and the movement of the fixed stars and the planets." For this reason Plato is to be considered an authority on the mathematical arts of the quadrivium's arithmetic, geometry, astronomy and music.[11] For William, by contrast, whose later commentary frequently cites the

8 For the traditional, 'Platonic' division, see Isidore of Seville, *Etymologiae* II.24.3–7, followed by Alcuin, *De dialectica* I (*PL* 101, 952C), and Rabanus Maurus, *De universo* V.1 (*PL* 111, 413–14). For Isidore's apparently unprecedented attribution of the quadrivium to Plato, see J. Fontaine, *Isidore de Seville et la culture classique dans l'Espagne wisigothique*, 3 vols. 2nd ed. (Paris: Études Augustiniennes, 1983), III: 1124–25.

9 *De trinitate* II, in Boethius, *De consolatione Philosophiae; Opuscula theologica*, ed. C. Moreschini (Leipzig: Teubner, 2000), 168–69: "Nam cum tres sint speculativae partes, naturalis, in motu inabstracta ἀνυπεξαίρετος (considerat enim corporum formas cum materia, quae a corporibus actu separari non possunt: quae corpora in motu sunt ut cum terra deorsum ignis sursum fertur, habetque motum forma materiae coniuncta) . . ." For twelfth-century discussions, see *Commentaries on Boethius by Thierry of Chartres and his School*, ed. Häring, 68–85, 154–77, 336–55, 540–44.

10 For the attribution, see Bernard of Chartres, *Glosae super Platonem*, ed. P. E. Dutton (Toronto: PIMS, 1991). For a critique, with bibliography of previous discussion, see P. Dronke, *The Spell of Calcidius: Platonic Concepts and Images in the Medieval West* (Florence: SISMEL, 2008), xxi, n. 39, 136–40.

11 Bernard of Chartres, *Glosae super Platonem*, ed. Dutton, 141.58–63: "Ad phisicam tendit, cum de planis figuris et solidis corporibus, de incorporatione animae mundi et aliarum, earumque motu perpetuo, de stellarum discursibus ratis et errantibus loquitur. Unde seruata omnium artium fere ratione, hoc opus non rudibus, sed in quadruuio promotis elaboratum est, ut si quae quaestiones de musica et aliis oriuntur, domesticis rationibus, scilicet musicis, arithmeticis, et ceteris, sopiantur." Here Bernard was following Calcidius, *Commentarius* II, ed. Waszink, 58.

medical authority of Constantine the African, *physica* concerns "the nature and complexion of bodies," and Plato's dialogue contributes to that science when it speaks of "the four elements, the creation of animals, and of primordial matter."[12] Under the influence of theoretical medicine, physics can be seen here emerging from the shadows of the quadrivium as an independent intellectual discipline. A parallel development may also be observed in the case of the astrologer Hermann of Carinthia, who, making use of Arabic divisions of philosophy, advanced a definition of *physica* in terms of celestial influence that would have been largely unfamiliar to Latin scholars in the 1140s.[13]

How did the *Cosmographia*, composed around 1147, relate to these recent developments in the study of nature and the definition of natural science? Opinions have varied about how best to characterise the relationship of Bernard's creation allegory to the more strictly discursive philosophy of contemporaries such as William of Conches and Hermann of Carinthia. One critic has judged the *Cosmographia* "somewhat older fashioned" in its dependence on traditional sources at a time of innovation in natural philosophy.[14] Another has written of Bernard's relationship to his immediate precursors that he "crystallises in a remarkably comprehensive form the aspirations and attainments of the cosmologists of the earlier period."[15] The identification of sources and influences on the *Cosmographia* obviously plays an important role in determining its position in twelfth-century discussion. However, this chapter will argue that the work's significance does not turn on the question of whether it

---

12 William of Conches, *Glosae super Platonem*, ed. É. Jeauneau, CCCM 203 (Turnhout: Brepols, 2006), 10. 18–19: "Phisica uero est de naturis et complexionibus corporum: phisis enim est natura"; and 11.7–8: "... ubi uero de quatuor elementis et creatione animalium et de primordiali materia, de phisica." See also the diagrammatic division of philosophy on p. 11 of Jeauneau's edition.

13 In Hermann's "Letter-Preface to the Translation of Ptolemy's *Planisphere*", in *De essentiis*, ed. Burnett, 347–49, he refers (348.27–31) to his *De essentiis* as a work in progress on the subject of *physica*, "cuius prior pars superioris mundi, ut sequens inferioris, naturam contemplatur—id autem est materiales rerum causas, quemadmodum illa formales, omnis videlicet geniture principia post primam ipsam causam utrumque moventem, ut in eo quod de essentiis instituimus, plenius patebit." On the Arabic division of philosophy, see J. Jolivet, "The Arabic Inheritance", in *History of Twelfth-Century Western Philosophy*, ed. Dronke, 113–48; and for the place of *physica* in such a division, see the *c.* 1150 treatise of the Toledan translator Dominicus Gundissalinus, *De divisione philosophiae*, ed. L. Baur (Münster: Aschendorff, 1903), 19–27.

14 T. Silverstein, "*Liber Hermetis Mercurii Triplicis de vi rerum principiis*", *AHDLMA* 22 (1955): 217–302 (237).

15 W. Wetherbee, "Philosophy, Cosmology and the Twelfth-Century Renaissance", in *History of Twelfth-Century Western Philosophy*, ed. Dronke, 21–53 (45).

was at the very avant-garde of mid-century philosophy. Rather, the importance of the *Cosmographia* will be seen to lie in its combination of the traditional techniques of literary *imitatio* with certain natural philosophical doctrines current in the scholarly circles in which Bernard worked. The product of this was a work unlike any other written in the period—one that revised a number of ancient authorities to offer a searching analysis of the purpose of natural science in the fallen world. To develop this argument it will first be useful to discuss the significance of the work's form.

The Mixed Form

The form of the *Cosmographia* alternates between sections of metre and prose. Reviving what was taken to be the standard medieval usage, modern scholarship has tended to call this kind of composition a *prosimetrum*.[16] However, Anne-Marie Turcan-Verkerk has demonstrated that in its earliest uses, the term referred not to a work that alternated verse and prose, but rather to a style that combined elements of rhythmical and quantitative metre.[17] The word *prosimetrum* first appears in the *Rationes dictandi* (1119–1124) of Hugh of Bologna, who proposed a threefold division of metrical composition (*dictamen metricum*):

> Hoc autem repperitur tripliciter: aut cum pedum mensura, et carmen uocatur; uel numero dumtaxat sillabarum cum uocum consonantia, et tunc riddimus appellatur; seu utroque mixtum, quod quidem prosimetrum conpositione dicitur, ut sequentia declarant exempla. Est enim carmen: *bella per Emathios*. Riddimus solus ut hic: *hostis Herodes impie*, et cetera, et est cantui satis congruus, quia utrobique repperitur octogonus. Ceterum prosimetrum possumus dicere, quando pars uersifice pars uero profertur prosaice, ut exempla declarant: *Ugo, patris matrisque loco quem habui semper, quicquid habet quecumque ualet dat mihi libenter*.[18]

16 The earliest use of the word in modern criticism (cited by Turcan-Verkerk, 112: see following note) is apparently in E. Norden, *Die antike Kunstprosa vom VI. Jahrhundert v. Chr. bis in die Zeit der Renaissance*, 2 vols. (Leipzig: Teubner, 1898), II: 756.

17 A.-M. Turcan-Verkerk, "Le *Prosimetrum* des *Artes dictaminis* médiévales (xiie–xiiie s.)", *Archivum Latinitatis Medii Aevi* 61 (2003): 111–74.

18 Hugh of Bologna, *Rationes dictandi prosaice*, ed. L. Rockinger, *Briefsteller und Formelbücher des eilften bis vierzehnten Jahrhunderts*, 2 vols. (Munich: Franz, 1863–1864), I: 47–94 (54–55).

> Metre occurs in three forms: in a measure of feet—and this is called a poem [*carmen*]—or in a specified number of syllables with harmonious accents—and then it is named a rhythm [*riddimus*]—or else it is a mixture of both, which is said by its composition to be a *prosimetrum*, as the following examples show. For this is a poem: *bēllă pĕr Ēmăthĭōs*. Rhythm on its own is like this: *hóstis Heródes ímpie* etc, and it is well suited to chant, because in both cases it has eight parts. We may call the remaining composition a *prosimetrum* when it is uttered partly in verse and partly in prose, as the examples show: *Ūgŏ, pătrīs mātrīsqŭe lŏcō quēm hábui sémper, quīcquĭd hăbēt quēcūmquĕ uălēt dāt míhi libénter*.

While Hugh's example of *carmen* is the first half of a hexameter, and his *riddimus* a sequence of stressed and unstressed syllables, the illustration of *prosimetrum* is more complex. The sentence is "a mixture of both" *carmen* and *riddimus* because it may be divided into two halves, each of which starts with the first four feet of a hexameter, before ending with rhyming phrases that switch from vowel quantities to stress accents. It is unclear whether Hugh considered the *prosimetrum* to be a metrical prose style, or a verse form that combined quantities and rhythm—for medieval authors sometimes referred to rhythm as *prosa*. The distinction may be an unhelpfully modern one.[19] But what is clear is that for Hugh the *prosimetrum* was a style of writing, not a genre in which prose and verse alternated. The term appears not to have been used at all in twelfth-century France, but Hugh's discussion was taken up by a number of his Italian followers, some of whom removed the reference to the mixture of quantitative and rhythmical elements—and even the illustrative quotation—defining the *prosimetrum* simply as a composition in which *pars metrice, pars prosaice scribitur*.[20] It was apparently only from the turn of the thirteenth century, and as a result of such truncated definitions, that dictaminists came to confuse the *prosimetrum* with the mixed form of Boethius's *De consolatione Philosophiae*, which commentators since the ninth century had similarly described as a work composed *partim prosa, partim metro*.[21]

If Bernard Silvestris did not think of his *Cosmographia* as a *prosimetrum*, would he have had another name for the form of the work? A possible alternative is the word *satura*, or satire. In 1978 Udo Kindermann argued that

---

19 See Turcan-Verkerk, "Le *Prosimetrum*", 117–19.

20 See the quotations from Adalbertus Samaritanus's *Praecepta dictaminum* and the *Traité Lombard* cited by Turcan-Verkerk, "Le *Prosimetrum*", 122, 124.

21 Turcan-Verkerk, "Le *Prosimetrum*", 132; and for the Boethius commentators, see Pabst, *Prosimetrum*, I: 284–307.

medieval scholars used this term to label works that combined verse and prose, following the lead of Martianus Capella, whose fifth-century verse and prose encyclopaedia *De nuptiis Philologiae et Mercurii* was apparently related to him by the muse Satura.[22] Bernhard Pabst, however, has shown that medieval commentators did not in fact see Martianus's muse as a symbol of the mixed form; when they did read Satura as a marker of genre, they saw her rather as a mocker of vice, as in traditional Roman satire.[23] The most promising association between *satura* and the mixed form appears in a ninth-century *Vita Boetii*, whose author says that Boethius imitated and surpassed Martianus Capella in his use of satire:

> Hos libros per satiram edidit imitatus uidelicet Marcianum Felicem Capellam qui prius libros de nuptiis Philologiae et Mercurii eadem specie poematis conscripserat, sed iste longe nobiliore materia et facundia ei praecellit quippe nec Tullio in prosa nec Virgilio in metro fuit inferior.[24]
>
> Boethius published these books of satire in imitation of Martianus Felix Capella, who had previously composed books *On the Marriage of Philology and Mercury* in the same kind of poetry; but Boethius far surpassed him in eloquence and nobility of subject matter, for he was the equal of Cicero in prose and Vergil in verse.

Even here, though, it is not entirely clear that the kind of poetry (*species poematis*) in question is the type that is mixed with prose. Isidore of Seville's definitions of satire linked it with the *lanx satura*—the plate laden for sacrifice with varied fruits—for in their efforts to ridicule vice, satirists would 'saturate' their verse with many topics at once: thus "to write satires is to compose varied poetry [*poemata varia*]."[25] Isidore gives as examples of *poemata varia* the

---

22 U. Kindermann, *Satyra. Die Theorie der Satire im Mittellateinischen: Vorstudie zu einer Gattungsgeschichte* (Nürnberg: Hans Karl, 1978), 21–22. See Martianus Capella, *De nuptiis Philologiae et Mercurii* I.2, ed. J. Willis (Leipzig: Teubner, 1983), 2.15–16: "… fabellam tibi, quam Satura comminiscens hiemali pervigilio marcescentes mecum lucernas edocuit, ni prolixitas perculerit, explicabo."

23 Pabst, *Prosimetrum*, I: 275–81.

24 *Vita Boetii* I, in Boethius, *Philosophiae consolationis libri quinque*, ed. R. Peiper (Leipzig: Teubner, 1871), xxxi. The *Accessus Boetii*, in *Accesses ad auctores. Bernard d'Utrecht. Conrad d'Hirsau, Dialogus super auctores*, ed. R. B. C. Huygens, 2nd ed. (Leiden: Brill, 1970), 47.11–16, has almost identical wording, and according to Huygens is based on the earlier *Vita*.

25 Isidore of Seville, *Etymologiae* V.16, ed. W. M. Lindsay, 2 vols. (Oxford: Clarendon Press, 1911): "saturas scribere est poemata varia condere, ut Horatii, Iuvenalis et Persii." See also

hexameter satires of Horace, Juvenal and Persius, but his definition seems to have been misunderstood by at least one Boethius commentator, who glossed the term *satira* in the above-mentioned *Vita Boetii* as meaning "a variety of metres", thereby linking satire with polymetry, rather than with the varied *content* of the monometric authors listed by Isidore.[26] Given that Martianus Capella employed fifteen different metres, and Boethius twenty-eight, it is at least conceivable that the author of the *Vita Boetii* also had this polymetric meaning in mind. At the very least then, the term *satura*, like *prosimetrum*, meant different things to different authors. For our purposes, the important point is that there is no twelfth-century evidence that either word was used to refer to the *Cosmographia*, or—to my knowledge—to similar works of prose and verse composed in the period. In what follows, then, I shall adopt the more neutral designation of 'the mixed form' to describe the prose and verse arrangement of such works.

Whatever word Bernard would have used to describe his *Cosmographia*, he was without doubt deeply influenced by Martianus Capella and Boethius, his two greatest precursors in the mixed form.[27] Bernard's debts to these writers are worth exploring for what they reveal about the distinctive scientific vision of his own treatise. Both Martianus and Boethius were curriculum authors, whose works had been objects of a tradition of glossing and interpretation stretching back to the ninth century.[28] In the more immediate past, the mixed

---

Etym. VIII.7.7–8: "Novi [comici], qui et Satirici, a quibus generaliter vitia carpuntur, ut Flaccus, Persius, Iuvenalis vel alii. Hi enim universorum delicta corripiunt, nec vitabatur eis pessimum quemque describere, nec cuilibet peccata moresque reprehendere. Vnde et nudi pinguntur, eo quod per eos vitia singula denudentur. Saturici autem dicti, sive quod pleni sint omni facundia, sive a saturitate et copia: de pluribus enim simul rebus loquuntur; seu ab illa lance quae diversis frugum vel pomorum generibus ad templa gentilium solebat deferri; aut a satyris nomen tractum, qui inulta habent ea quae per vinolentiam dicuntur."

26 *Anonymus in Boetii Consolationem Philosophiae Commentarius ex Sangallensis Codice, Liber Primus*, ed. G. C. Roti (Doctoral Dissertation: State University of New York at Albany, 1979), 3: "*Per satiram*: per poetriam, id est per varietatem metrorum. Satiricus dicitur reprehensor viciorum."

27 For Bernard's debts to them, see Pabst, *Prosimetrum*, I: 446–76.

28 For a survey of the early medieval reception of Martianus, see M. Teeuwen, *Harmony and the Music of the Spheres: The Ars Musica in Ninth-Century Commentaries on Martianus Capella* (Leiden: Brill, 2000), and also her article "The Pursuit of Secular Learning: The Oldest Commentary Tradition on Martianus Capella", *The Journal of Medieval Latin* 18 (2008): 36–51, which notes the possibility that the oldest commentary tradition may be traced to Tours in the 820s–830s. For Boethius, see J. Beaumont, "The Latin Tradition of the *De consolatione Philosophiae*", in *Boethius: His Life, Thought, and Influence*, ed.

form of these writings had been revived by two other works with connections to Tours: the *De querimonia carnis et spiritus* (*c.* 1101) of Hildebert of Lavardin, subsequently Archbishop of Tours (1125–1133), and the *De eodem et diverso* (1104 or 1109–1116) of Adelard of Bath, who had studied in the city shortly before writing his treatise. In a recent study, Bridget Balint has shown how Bernard and these precursors shared "an overriding concern with the place of humanity within a divinely ordered cosmos", as they developed and refined their ancient models to respond to contemporary intellectual concerns.[29] As we have seen, it was a commonplace of medieval commentators that Boethius had imitated and surpassed Martianus Capella.[30] In a similar way, Balint has argued that Hildebert of Lavardin's *De querimonia* aimed "to imitate and supersede" and "to update" the *De consolatione* by parodying the relationship between narrator and authority figure (Boethius's Philosophia/Hildebert's Anima).[31] The *Cosmographia* likewise continued this pattern of imitation and refinement of its ancient models, in ways that are revealing of the purpose of the work and its relationship to the intellectual climate in which it was written. The *Cosmographia*'s debts to, and revisions of, Boethius will be treated later in this chapter, but here a comparison with Martianus Capella will help to clarify the scope of Bernard's project.

In the *De nuptiis Philologiae et Mercurii*, the earthly maiden Philology is summoned to the stars as a worthy bride for Mercury, on account of her great learning. According to the Neoplatonic suppositions of the work, this apotheosis represents the soul's return to its birthplace in the stars, and in the allegory of the first two books it is suggested that this apotheosis is achieved through study. The crucial role of philosophy in the soul's return to its celestial home is symbolised by Philology's wedding-gift of seven maidens representing the seven arts, the content of which is summarised in the remaining seven encyclopaedic books of the work. For the scholars who rediscovered and popularised the *De nuptiis* in the ninth century, Martianus offered a myth of knowledge-to-be-recollected that was compatible with their Christian ideas

---

M. Gibson (Oxford: Blackwell, 1981), 278–305, supplemented by F. Troncarelli, *Boethiana Aetas: modelli grafici e fortuna manoscritta della "Consolatio Philosophiae" tra ix e xii secolo* (Alessandria: Edizioni dell'orso, 1987).

29 Balint, *Ordering Chaos*, 17.

30 In addition to the texts cited above (n. 24), see those assembled by Pabst, *Prosimetrum*, I: 284.

31 Balint, *Ordering Chaos*, 40, 24. Still valuable for its more extended treatment of a neglected work is Balint's dissertation: "Hildebert of Lavardin's *Liber de querimonia* in its Cultural Context" (Doctoral Dissertation: Harvard University, 2002).

about the restoration of the soul from original sin.[32] In particular, the image of the seven maidens proved to be a rich source of inspiration for literary and pictorial allegories of the seven arts from the ninth through to the twelfth century.[33] It appears, for instance, as a prominent narrative device in Adelard of Bath's *De eodem et diverso*. A treatise on the liberal arts, written in the mixed form and dating from the first two decades of the twelfth century, this work is an allegorical account of Adelard's education in Tours before he went in search of new scientific learning in southern Italy and Sicily—presumably where he wrote the *De eodem et diverso*, which is dedicated to William, Bishop of Syracuse. As in the *De nuptiis*, the seven arts are allegorised as a succession of maidens whose gifts to the scholar are summarised by Philosophia, who states that they provide the soul with a means "by which she can restore herself to herself and bring herself back home."[34]

The liberal arts content of Adelard's curriculum was entirely traditional.[35] It is only at the end of the work that he alludes to the study of natural science, not within the framework of the arts allegory, but as an anecdotal coda to the treatise. There Adelard recalls a conversation he had in Southern Italy about magnetism, with a man who "more than anything else could talk about the art of medicine and the nature of things."[36] This anecdote reveals Adelard's

32 J. Contreni, "John Scottus, Martin Hiberniensis, the Liberal Arts, and Teaching", in *Insular Latin Studies: Papers on Latin Texts and Manuscripts of the British Isles, 550–1066*, ed. M. W. Herren (Toronto: PIMS, 1981), 23–44 (repr. in Contreni, *Carolingian Learning, Masters and Manuscripts* (Aldershot: Ashgate, 1992), VI); C. E. Lutz, "Remigius' Ideas on the Classification of the Seven Liberal Arts", *Traditio* 12 (1956): 65–86.

33 D'Alverny, "La Sagesse".

34 *De eodem et diverso*, in Adelard of Bath, *Conversations with his Nephew*, ed. and trans. Burnett *et al.*, 32: "His igitur anima in corporis carcere vinculis oppressa, unum inter universa remedium est quo eadem se sibi reddit domumque reducit—doctrine videlicet huius philosophie, artesque quas vocant liberales." Compare the discussion of reading as *anagnosis* (knowledge regained) in Macrobius, *Commentarii in somnium Scipionis* I.21.1, ed. Willis, 49.25–29: "Hinc est quod quae apud Latinos lectio, apud Graecos vocatur repetita cognitio, quia cum vera discimus, ea recognoscimus quae naturaliter noveramus, prius quam materialis influxio in corpus venientes animas ebriaret."

35 A. Drew, "*De eodem et diverso*", in *Adelard of Bath: An English Scientist and Arabist of the Early Twelfth Century*, ed. C. Burnett (London: Warburg Institute, 1987), 17–24.

36 *De eodem et diverso*, 70: "Et ego certe cum a Salerno veniens in Grecia Maiore quendam philosophum Grecum qui pre ceteris artem medicine naturasque rerum disserebat, sententiis pretemptarem, causam scilicet querens qua vi et natura magnetes ad se ferrum trahant, eiusque super hac re ceterisque similibus solutione audita, ipse michi vicissim proponeret, dicens…" (I have corrected the editors' *trahat* to *trahant*.) C. Burnett, "Physics before the *Physics*: Early Translations from Arabic of Texts Concerning Nature in

developing interest in natural scientific questions, but it suggests that such questions were beyond the scope of the northern French curriculum around the turn of the twelfth century. They belonged to what Adelard referred to elsewhere as the *studia Arabum* as opposed to the *sententiae Gallicae*.[37] In the *De eodem et diverso*, medicine and the study of *naturae rerum* are presented as an appendage to the arts, sitting in an unclear relationship to them.

Bernard's work shares many themes with those of Martianus and Adelard. In the Neoplatonic narrative of the *Cosmographia*, the human soul is created in the heavens, whence it descends through the cosmos into a state of corporeal *alteritas* (II.8.31). After death the soul will return to its home in the stars (II.10.51–53), but for the term of its terrestrial life it is condemned to inhabit a bodily form that Bernard likens to a prison or tomb.[38] What the soul forgets in its incarceration is the knowledge of the elements, planets and stars instilled by Urania during its descent through the cosmos. This narrative of cosmic descent is "a type of *Bildungsroman*", as Brian Stock has nicely put it.[39] Although the knowledge instilled on this journey is obscured by corporeality, it may nonetheless be recollected in the present life. In her summary of what the mind will learn from her during its cosmic descent, Urania declares:

> More recordantis quam multa reducet eorum
> Que cernet, penitus non immemor. (II.4.37–38)
>
> Like a man remembering, how much the mind will recall,
> Being not entirely forgetful!

As a description of the soul's descent and a summary of its education by Urania, the *Cosmographia*, like the *De nuptiis* and Adelard's *De eodem et diverso* before it, is designed to stimulate the recollection of lost knowledge. Unlike those authorities though, Bernard did not focus on the traditional curriculum of the seven arts. He makes occasional reference to an exemplary figure in

MSS British Library, Additional 22719 and Cotton Galba E IV", *Medioevo* 27 (2002): 53–109, edits some texts relating to the kind of discussions *de naturis rerum* in southern Italy to which Adelard may refer.

37 *Questiones naturales*, in Adelard of Bath, *Conversations with his Nephew*, ed. and trans. Burnett *et al.*, 90.

38 *Cos.* II.8.33–34: "Sed que compedibus, que carcere clausa tenetur, | En quasi corporea mole sepulta iacet!"

39 Stock, *Myth and Science*, 164. Likewise Pabst, *Prosimetrum*, I: 459: "[eine] stufenweise Bildungsreise".

rhetoric or geometry ("Et Cicero rhetor, et geometra Thales", I.3.46), or to the proficiency of Gaul in logic ("Quicquid ad elencos arguto disputat ore | Gallus", II.14.75–76), but beyond these casual examples there is no sustained treatment of the seven arts, nor any argument for their conceptual importance for the soul's self-recollection in the fallen world. The one exception is the science of the stars, which Bernard treats extensively, and which relates to his project, as stated in the dedication to Thierry of Chartres, of writing a treatise "de mundo, de universitate."[40] The study of nature, then, which was still only a marginal presence three decades before in Adelard's *De eodem et diverso*, has come to occupy a central position in the *Cosmographia*, as the knowledge that may restore man to his original dignity.

Platonic Themes and Variations

"In causas rerum sentit Plato", Bernard wrote (I.3.51), and it is to Plato's *Timaeus* that the natural philosophy of the *Cosmographia* is most deeply indebted.[41] Brought into circulation in the French schools during the late eighth century, Calcidius's partial fourth-century translation of the *Timaeus*, increasingly read independently of his commentary, survives in around fifty twelfth-century manuscripts and was well known to Bernard's contemporaries.[42] Like Plato, Bernard begins with a beneficent deity who desires to fashion the best possible world.[43] Bernard calls his deity Noys but also *opifex* (*Cos.* I.3.57), one of the words that Calcidius used to translate Plato's *demiourgos*. In both works this creative deity models the forms of life on its own ideas.[44] The physical construction of the world begins with the ordering of the elements, a process

40 *Cos.*, 96.5–7: "Siquidem de mundo, de universitate tractatus sua natura difficilis, sed et sensu tardiore conpositus, sicut aures sic oculos arguti iudicis reformidat."

41 Tauste Alcocer, *Opus naturae*, 391–99. For discussion of Bernard's use of the innovative philosophical language of Calcidius, see Dronke, *Spell of Calcidius*, 8–12.

42 A. Somfai, "The Eleventh-Century Shift in the Reception of Plato's *Timaeus* and Calcidius's *Commentary*", *Journal of the Warburg and Courtauld Institutes* 65 (2002): 1–21.

43 *Cos.* I.1.11–14: "Nempe deus, cuius summe natura benigna est, | Larga, nec invidie miseros sensura tumultus, | In melius, quantum patitur substantia rerum, | Cuncta refert, operique suo non derogat auctor."
Compare *Timaeus* 29E, ed. Waszink, 22.18–20: "Optimus erat, ab optimo porro invidia longe relegata est. Itaque consequentur cuncta sui similia, prout cuiusque natura capax beatitudinis esse poterat, effici voluit."

44 *Cos.* I.2.8.4–5: "[Noys], ad eternas introspiciens nostiones, germana et proximante similitudine rerum species reformavit."

described by both philosophers (*Cos.* I.2.4–11; *Timaeus* 32BD), and to this material is added the enlivening impulses of the *anima mundi*, a term that Bernard borrowed from Plato as an alternative name for his character Endelechia.[45] After the interlude of the catalogue poem, Bernard concludes the "Megacosmus" with some reflections on the relationship between the sensible universe and its intelligible model that take their cue from Plato's dictum that time is a moving image of eternity.[46] In the second book of the *Cosmographia*, as in the *Timaeus*, man is constructed as a microcosm, in whom the world soul and the elemental leftovers from creation are combined.[47] The structure of his body also conforms to the model of the universe, with man's rounded head being

---

Compare *Timaeus* 30CD, 23.15–20: "... siquidem animalium genera mundus alter complectitur perinde ut hic nos et cetera subiecta visui et ceteris sensibus. Ergo intellegibili substantiae praecellenti principalique naturae omnifariam quoque perfectae deus opifex gigni simile volens sensibile animal unum et visibile constituit, naturae suae convenientia cuncta quae vita fruuntur intra conseptum et limitem suum continens."

45 Bernard would have found the term Endelechia in Martianus Capella (*De nuptiis* I.7, II.213) and more pertinently in Calcidius (*Commentarius* CCXXII), who, following Aristotle, defines *endelechia* as the actualisation of matter that had previously existed in potential, which is exactly the role played by Endelechia/Anima mundi in the *Cosmographia*. In Calcidius, then, and in Bernard following him, this term seems not in fact to denote the Greek ἐνδελέχεια (continuance) but rather—with a τ not a δ—ἐντελέχεια, that is the 'actualisation' of matter. On this see Dronke, *Spell of Calcidius*, 71–74, who adds (71, n. 1) that "in the *De nuptiis* (I.7 and II.213) Martianus, like Calcidius, was writing about entelechy and not about continuance"—an interpretation shared by Martianus's modern editors, who conjecturally emend the *endelechia* of the manuscripts, but contested by D. Shanzer, *A Philosophical and Literary Commentary on Martianus Capella's "De nuptiis Philologiae et Mercurii", Book I* (Berkeley: University of California Press, 1986), 68.

46 *Cos.* I.4.3.3–7: "Ex eo incipientis vite primordio, cum volvente celo de motu quoque siderum substantia temporis nasceretur, que successerunt secula—simplici eternitatis initiata principio—cum sua numerus varietate suscepit"; I.4.13.1–2: "Eternitas igitur, sed et eternitatis imago tempus, in moderando mundo curam et operam partiuntur."
Compare *Timaeus* 37D, 30.2–4: "Quapropter imaginem eius mobilem numeroque serpentem factae a se machinae deus sociabat eam quae tempus dicitur, aevo intacto et in singularitate perseverante."

47 *Cos.* II.11.2.2–4 (Noys, addressing Urania, Natura and Physis): "In vestra manu est sumere, cum velitis, humane sementem anime, ex Endelichia iam creata; corpus, ex elementis massa confusionis exceptis."
Compare *Timaeus* 41D, 36.14–16: "Haec dixit [opifex] et demum reliquias prioris concretionis, ex qua mundi animam commiscuerat, in eiusdem crateris sinum refundens eodem propemodum genere atque eadem ratione miscebat ..."

an image of the firmament.[48] The head is the seat of reason, and it is through the activity of rational thought—that is the pursuit of the good life—that the human *anima* may finally leave the temporary enclosure of the body and return to its home among the stars.[49]

For all his dependence on Plato for the structure, themes and wording of his treatise, Bernard made at least one major revision to this *auctor* that is revealing of the distinctive scientific outlook of the *Cosmographia*. In the *Timaeus*, the material world is not brought into being directly by the divine craftsman, the *opifex*, but rather by "sons of the gods"—the *dii deorum*.[50] Because the *opifex* may create only that which is incorruptible, he has to depute to lesser beings the ordering of the mutable cosmos. There were, then, two stages to creation: first, the *opifex* created the forms, elements and souls; then the lower powers, the *dii deorum*, united these into mutable bodies. In the *Cosmographia* the anonymous *dii deorum* are replaced by three carefully characterised goddesses: Urania, Natura and Physis.[51] These deities are the daughters of Noys, brought into being, their mother tells them, before the beginning of time.[52]

Urania is named, through Macrobius and Martianus Capella, after the eighth muse of Hesiod (*Theogony*, l. 78), whose name *ourania* (heavenly one) denotes her place among the stars.[53] Bernard calls her a "starry queen" ("reginam sideream", II.3.6.3) and the "native companion of the heavens" ("assis-

48 *Cos.* II.13.11.4–7: "Loco namque principe, de firmamenti spereque superioris exemplo, capud edificat in rotundo: capud tanquam arcem, tanquam tocius corporis capitolium, tollit et erigit in excelso."
Compare *Timaeus* 44D, 40.19–22: "Principio figuram capitis divinae potestates, quibus informandi corporis erat officium concreditum ex mundi figura mutuatae teretem globosamque finxerunt eidemque duos circuitus venerandae divinitatis innexuerunt."

49 *Cos.* II.8.33–36: "Sed que compedibus, que carcere clausa tenetur, | En quasi corporea mole sepulta iacet! | Ad natale iubar, ad regna paterna redibit, | Si sapiat; si non—coniuga carnis eat."
Compare *Timaeus* 42B, 37.9–11: "Et victricibus quidem ad comparis stellae contubernium sedemque reditum patere acturis deinceps vitam veram et beatam ..."

50 *Timaeus* 41A, 35.9–10.

51 *Cos.* II.11.1.1–6 (Noys speaking): "Trina igitur tribus superincunbit opera, cuique sua. Compositio anime: ex Endelichia et virtutum edificatione; corporis: ex materie preparatione; utrorumque corporis et anime formativa concrecio: de celestis ordinis emulatione. Prior igitur ad Uraniam, secunda ad Physim, tertia ad te, o Natura, dinoscitur pertinere." I have repunctuated this passage according to the suggestions of Minkova, "Bernardus Silvestris's *Cosmographia*", 161.

52 *Cos.* II.10.1–2: "Pignora cara, dee, quas ante creata creavi | Secula."

53 Macrobius, *Commentarii* II.3.2, ed. Willis, 104.3–6: "Unde Hesiodus in Theogonia sua octavam Musam Uraniam vocat, quia post septem vagas, quae subiectae sunt, octava stellifera

tricem indigentemque celi", II.3.12.3–4). She dwells in close proximity to the throne of Noys (II.3.3.9), in the fixed stars of the Aplanon, at the outer limit of the universe (II.3.9.8–9), and her domain descends from there to the border of the moon (II.9.1.4–5). Urania is portrayed as being expert in both astronomy and astrology, calculating the orbits of celestial bodies "according to sure standards of observation" ("ad certas observationis regulas", II.3.12.6), and from these discerning what the laws of fate permit or deny ("Quid fati series detve negetve", II.8.16). Thus to the purely astronomical identity of Urania in Bernard's Neoplatonic sources are added further astrological capabilities.

The domain of Physis begins where that of Urania ends, extending from the lunar boundary down to the "flourishing lap of the Earth" ("florentis terre gremio", II.9.1.12), where she resides among the lesser creatures ("Phisim in inferioribus reperies conversantem", II.3.3.9–10). Like Urania, Physis inspects the stars, but in order to explain how their composition affects the complexions of animals such as the hare, lion, fox and donkey.[54] She also creates remedies to correct the imbalances caused by disease.[55] This combination of theoretical physiology on the one hand, and a pharmacopoeia of plants, metals and stones on the other, is one that Bernard allegorises through the daughters of Physis, whom he names Theorica and Practica.

Finally, the activities of Natura are not restricted to a specific portion of the universe, but extend from the upper firmament, where she finds Urania, to the bosom of the Earth, where she encounters Physis. As Urania tells Natura when they meet: "For truly, you and I are of one kind; your domain is the universe at large, mine this one portion of the ether."[56] Lexically, Physis and Natura should be entirely identical, since as countless Latin glossators had noted, *physis grece*

---

sphaera superposita proprio nomine caelum vocatur." Likewise in Martianus Capella, *De nuptiis* II.118, Urania is the muse of astronomy.

54 *Cos.* II.9.6.9–13: "Siquando celo vel sideribus inferius excogitavit, de complexione mores animalium studuit iudicare: pavitationem leporis ex frigido, leonis audaciam ex ignito, in vulpe versuciam, tarditatem in asino, alterum flegmatis, alterum melancolie qualitatibus importari."

55 *Cos.* II.9.7.4–13: "Ea contra inveniendis invigilabat remediis, ad quorum temperamentum inequalitas frenaretur, cum se dirigeret ad nocendum. Verum id disquisitiori pertractabat ingenio, ut elementa—partes mundi primarie partesque partium—in hiis, ex hiis, et per hec que generant, causis phisicalibus deservirent: non herbis, non plantis, non contenta graminibus, de lapidibus, de metallis remediales etiam extorqueret effectus. Id sue visum est superaddendum scientie, ut per mixturam calidam venenis etiam mortiferis ad curam medicam salubriter uteretur."

56 *Cos.* II.4.17–18: "Nempe tuum genus unde meum: te publicus orbis, | Unus me recipit locus etheris."

natura.[57] Bernard himself considered the two terms to be closely related, writing that Physis "followed Natura and whatever is included under that name."[58] Nonetheless, he took the apparently unprecedented step of making a formal distinction between the two. For Bernard, Physis represents the study of medicine informed by natural science, while Natura stands for a broader conception of the study of nature—she is simply the "rerum prudentior indagatrix" (II.7.1.7)—and one who encompasses and unites both medicine and astrology.

These three are the goddesses who create man, so an understanding of their capabilities is essential for understanding the capabilities of human life as portrayed in the *Cosmographia*. The habitual positions of the goddesses and their characteristic range of activities evoke a broader conception of natural philosophy than that transmitted by Bernard's late antique Platonic sources. The depiction of Urania, for instance, suggests the influence of those Arabic astrological texts that were starting to be studied in the French schools. Likewise the interest of Physis in complexions, and the inclusion of her daughters Theorica and Practica, suggest the theoretical and practical division of medicine introduced by the *Pantegni* and *Isagoge Iohannicii ad artem Galieni*, Arabic medical texts rendered into Latin in the eleventh century.[59] And finally, the distinction between Physis and Natura may correspond to a well-attested semantic shift, coinciding with the reception of the theoretical medicine of the southern Italian translations, whereby medicine came increasingly to be referred to as *physica*, instead of the less theoretically sophisticated *medicina* of the early Middle Ages.[60] This identification of Urania, Natura and Physis

57 *Novum Glossarium Mediae Latinitatis*, fasc. "Phacoides—Pingo" (Brussels: UAI, 2003), 120–21.

58 *Cos*. II.9.6.7–8: "... naturam, et quicquid eo nomine continetur ... sequebatur."

59 It would appear to be in this sense that Bernard wished Theorica and Practica to be understood rather than in the wider sense of the two principal branches of theoretical and practical knowledge in the Aristotelian division of philosophy. In Boethius's discussion of this division (see *In Isagogen Porphyrii commenta*, ed. S. Brandt (Vienna: Tempsky, 1906), 8–9), theoretical philosophy is subdivided into *intellectibilis* (metaphysics), *intellegibilis* (mathematics), and *naturalis* (physics), and it is only the third of these subdivisions "quae est physiologia, quae naturas corporum passionesque declarat" (9.7–8).

60 P. O. Kristeller, "The School of Salerno: Its Development and its Contribution to the History of Learning", *Bulletin of the History of Medicine* 17 (1945): 138–94; J. J. Bylebyl, "The Medical Meaning of *Physica*", *Osiris*, 2nd ser. 6 (1990): 16–41; M. Teeuwen, *The Vocabulary of Intellectual Life in the Middle Ages* (Turnhout: Brepols, 2003), 400–403. For three contemporary sources, see Thierry of Chartres, *Commentum super Boetii librum de trinitate* II.9–10, in *Commentaries on Boethius*, ed. Häring, 71, and the discussion of *physica theorica*

as branches of twelfth-century *scientia naturalis* has been widely accepted.[61] More controversial has been the extent of Bernard's use of the recent translations of Arabic medical and astrological texts that lie behind these generalities. A brief reappraisal of this debate will help to clarify the function of this natural philosophical material in the work.

Medicine and Astrology

In terms of medical sources, it has at one extreme been claimed that Bernard had access to some of the most recent works produced in southern Italy. There is, however, no compelling evidence to support such a view.[62] At the other extreme, Heinrich Schipperges has expressed doubt about any influence of the new Constantinian medicine on the *Cosmographia*.[63] It is indeed the case that many of the physiological doctrines in the work may be traced back to Platonic sources. Although the *Timaeus* did not provide much material for this aspect of cosmogony—the majority of Plato's physiology falling outside that part of the dialogue known through the Latin translations—*Timaeus* 44D–47E does contain a discussion of the human organs and senses, and Bernard's lines on the sense of sight clearly draw on the language of this passage.[64] But Bernard could also turn to other Latin sources to supplement the *Timaeus*. He appears, for instance, to have used Apuleius's *De Platone et eius dogmate* for his charac-

(cosmology) and *physica practica* (medicine) in John of Salisbury, *Policraticus* II.29, ed. Keats-Rohan, 169–71, and *Metalogicon* II.15, ed. Hall and Keats-Rohan, 78–79.

61 *The "Cosmographia" of Bernardus Silvestris*, trans. W. Wetherbee (New York: Columbia University Press, 1973), 43; Stock, *Myth and Science*, 195, 221; Lemay, *Abu Ma'shar*, 276–77.

62 For this view, see J. Jolivet, "Les principes féminins dans la *Cosmographia* de Bernard Silvestre", in *L'homme et son univers au Moyen Âge*, ed. C. Wenin, 2 vols. (Louvain: Institut Supérieur de Philosophie, 1986), I: 296–305, and for some problems with it, see below, Chapter 3, 107–8.

63 H. Schipperges, "Einflüsse arabischer Medizin auf die Mikrokosmosliteratur des 12. Jahrhunderts", in *Antike und Orient im Mittelalter*, ed. P. Wilpert and W. P. Eckert (Berlin: Walter de Gruyter, 1971), 129–53 (136–37).

64 *Cos.* II.14.15–22: "*Intima* lux animeque dies *concurrit* ad ignis | Solaris radios ethereumque *iubar*, | De quo concursu vis et natura videndi | Essendi causam materiamque trahit. | Ad formas rerum se porrigit huius *acumen* | Lucis, et examen iudiciale facit; | Iudicioque tamen non omnia sentit eodem: | Illius interdum languet, habundat opus." The italicised terms (and "domestica lux", l. 25) appear at *Timaeus* 45BC, ed. Waszink, 41.15–22.

terisation of the sensory organs as *machinamenta* and *internuntia sentiendi*.[65] And, across a range of topics, he relied on the physiological amplifications of the *Timaeus* supplied by Calcidius—as in the case, to cite just one example, of his lines on the soul's simplicity but at the same time its adaptability to the variety of data that it receives through the senses.[66]

However, the influence of Constantinian medicine may be clearly discerned at a number of points. In a reference to the pre-eminence of Italy in medical studies, Bernard refers to the authority of "whatever the Ligurian pronounces in the art of medicine."[67] And it would indeed be strange, two decades after William of Conches had berated his contemporaries for attending insufficiently to the writings of Constantine the African and other *physici*, if Bernard's physiology did not bear some traces of this material.[68] Nonetheless, it should be noted that in the two obvious parallels between the *Cosmographia* and Constantine's *Pantegni*, the relevant doctrines also appear in William of Conches's *Philosophia* (1125–1130). One of these is a relatively minor statement about the eye: following Constantine (and William) Bernard described the eye as surrounded by seven membranes, as opposed to Calcidius's four, and existing in a pair, so that if one should fail the other may continue to function.[69]

---

65 *Cos*. II.13.14.1–3: "Sic igitur ibidem, erga regiam capitis, machinamenta collocat internuntia sentiendi, ut, de proximo sensibus interpellantibus, de proximo prodeat qui iudicet intellectus." Compare *De Platone* I.13, in Apuleius, *De philosophia libri*, ed. C. Moreschini (Stuttgart: Teubner, 1991), 104.5–6, where *machinamenta* are defined as those things "quibus ad sentiendas et diiudicandas qualitates sensus instructi sunt"; and *De Platone* I.16, ed. Moreschini, 197.8.

66 *Cos*. II.13.15.1–7: "Verum simplex simplicis, unaque unius, anime virtus est, sed non uniformis egreditur. Per oculos enim visus, per aures egrediens est auditus. Ad eundem modum deformatur in ceteris, secundum diversitatem membrorum quibus utitur instrumentis. Sensus igitur uno de fonte prodeunt, sed diversas expediunt actiones. Ille colores, ille sonos, alter gustus, alter odoramina persentiscit." This is a conflation of Calcidius, *Commentarius* CCXX, ed. Waszink, 232.19–233.9 (on the Stoics), and *Commentarius* CCXIV, 292.13–17 (on the Atomists). For other Calcidian parallels in physiology, see McCrimmon, "Classical Philosophical Sources", 142–45; and *Cosmographia*, trans. Wetherbee, 163–64.

67 *Cos*. II.14.75–76: "Quicquid ... in medica iactitat arte Ligus." *Ligus* is presumably a metonym for 'Italian', but a surprising one given its northern association.

68 See William of Conches, *Philosophia* I.23, ed. G. Maurach, with the assistance of H. Telle (Pretoria: University of South Africa, 1980), 28: "Sunt quidam, qui neque Constantini scripta neque alterius physici umquam legerunt, ex superbia ab aliquo discere dedignantes, ex arrogantia quae nesciunt confingentes ..."

69 *Cos*. II.14.33–34, 37–38: "Neve lacessiri queat ut libet, intus adacta | Tegmine septeno lux tunicata latet ... Non nichil est hos esse duos: si langueat alter, | Ut vice participis suppleat alter opus." Compare Constantine the African, *De communibus medico cognitu*

A more substantial parallel may be found in Bernard's complexion theory. When, in the penultimate chapter of the *Cosmographia*, Physis comes to create the body of man, she must first draw forth from matter the qualities of moist, dry, warm and cold, and then combine these into pure elements, each of which is a mixture of two qualities (II.13.5–6). Thus far her work on the microcosm reflects the macrocosmic arrangements undertaken by Noys at the beginning of the work. However, because Physis must fashion a terrestrial rather than a celestial being, the elements cannot be used in their pure condition, but must be compounded into complexions, the elements of which are known as humours (II.13.7–8). The idea of the humours was already present in early medieval medical theory, but Bernard's reference to *complexiones* as the compositional principles of human physiology reflects a terminological innovation introduced by Constantine the African.[70] And his insistence that humans are unique in their humoural balance, compared to the imbalanced mixture of humours in animals, is likewise paralleled by Constantine's *Pantegni*.[71] Once again, though, both these notions may be found in William's *Philosophia*.[72] In both cases, then, it is possible that

---

necessariis locis [= *Pantegni theorica*] II.13, in *Opera*, 2 vols. (Basel: Heinrich Petri, 1539), II: 59: "Hii duo sunt ut forte uno patiente, alter vicem suam possit explere. Sunt autem uniuscuiusque decem efficientes causae, scilicet humores et septem tunicae"; and Calcidius, *Commentarius* CCXLVI, ed. Waszink, 257.12–13: "Oculi porro ipsius continentiam in quattuor membranis seu tunicis notaverunt disparili soliditate." Compare William of Conches, *Philosophia* IV.39, ed. Maurach, 107 (*de oculo*): "Oculos igitur est quaedam orbiculata substantia et clara, sed in superficie aliquantulum plana, ex tribus humoribus et septem tunicis constans ... Duos vero oculos natura constituit, ut si unus laederetur, alter remaneret."

70 *Cos.* II.13.7.2–7: "Sic igitur de contiguo et proximo cum se sibi traderent elementa, coherencia—que in animali conplexio dicitur—est inventa. Complexiones igitur elementarias humane sic applicat edificationi ut, quibus innascuntur principiis, eorum origines consequantur." See Constantine the African, *Pantegni theorica* I.4, in *Opera*, II: 7–8; and D. Jacquart, "De *crasis* à *complexio*: note sur le vocabulaire du tempérament en latin médiéval", in *Mémoires, V: textes médicaux latins antiques*, ed. G. Sabbah (Saint-Étienne: L'Université de Saint-Étienne, 1984), 71–76 (repr. in Jacquart, *Science médicale entre deux renaissances* (*xii*^{e} *s.–xv*^{e} *s.*) (Aldershot: Ashgate, 1997), VI).

71 *Cos.* II.13.8.2–7: "Intemperans enim humorum coherentia brutorum conplexionem sepius adsolet depravare. Asinus hebes est ex flegmate, leo iracundus ex colera, canis aerio totus inficitur odoratu. Sola et singularis, hominum conditio de humorum conplexu facta est in qualitatibus et quantitatibus temperato." Compare *Pantegni theorica* I.4, in *Opera*, II: 8: "Homo enim omnibus temperatior est animalis speciebus."

72 William of Conches, *Philosophia* IV.30, ed. Maurach, 103 (*de complexionibus*): "Nos vero dicimus nullum esse hominem, qui calidus non sit et humidus, sed tamen quidam plus, quidam minus. Verbi gratia, homo naturaliter calidus est et humidus et inter quattuor qualitates temperatus."

Bernard was working not directly from Constantine's bulky encyclopaedia, but rather from William's convenient digest, which survives in around twenty-four twelfth-century manuscripts.[73] This theory would fit well with what we know of Bernard's position at Tours, a city not known to have been at the forefront of the reception of the new science, and which may have lacked copies of some of the scientific texts available in other northern French schools. Bernard may instead have worked from a widely available 'secondary' source such as William's *Philosophia*, or else have remembered ideas on which he had attended lectures or discussed with other philosophers.

A similar situation obtains with regard to Bernard's use of astrological sources. In 1948, Theodore Silverstein was the first scholar to suggest that Bernard was drawing on some of the most recent Arabic-Latin astrology, specifically Abu Ma'shar's *Introductorium maius ad scientiam iudiciorum astrorum* (translated in 1133 by John of Seville, and in 1140 by Hermann of Carinthia), and Hermann's own cosmological treatise, *De essentiis* (1143).[74] In 1962, Richard Lemay developed this thesis further, arguing that Abu Ma'shar and Hermann were in fact Bernard's "habitual sources", and that their works provided the key to unlocking the fundamentally astrological meaning of the *Cosmographia*.[75] Brian Stock followed these suggestions, confidently in the case of Abu Ma'shar, and more tentatively for Hermann of Carinthia, but like Silverstein he insisted on a wider range of sources for Bernard's cosmology than Lemay had posited.[76] The results of these source-critical studies are apparent in the notes to Winthrop Wetherbee's 1973 translation of the *Cosmographia*, which include eight references to Abu Ma'shar and fourteen to Hermann.[77] More recently though, there has been a reassessment of the significance of Bernard's ideas about celestial influence and the extent of his dependence on newly available Arabic astrology. Jim Tester has argued that "there is much less astrology in this strange work than might at first appear", and Charles Burnett that Lemay's "case for Bernard's indebtedness to Abu Ma'shar is not convincing".[78] Which of these is the correct position?

---

73 For the manuscripts, see P. E. Dutton, *The Mystery of the Missing Heresy Trial of William of Conches* (Toronto: PIMS, 2006), 37–40.

74 Silverstein, "Fabulous Cosmogony", 96–97.

75 Lemay, *Abu Ma'shar*, 258–84 (282).

76 Stock, *Myth and Science*, 27 (Abu Ma'shar), 113, n. 76 (Hermann of Carinthia).

77 *Cosmographia*, trans. Wetherbee, 174, 177.

78 J. Tester, *A History of Western Astrology* (Woodridge, Suff.: Boydell and Brewer, 1987), 145 (and n. 82: "there is less astrology in the work than Stock has seen there"); Burnett, "Establishment of Medieval Hermeticism", 128, n. 35. See also the sceptical remarks of Ratkowitsch, *Cosmographia*, 55–56.

Lemay pointed to five specific passages where he believed Bernard to have borrowed from Abu Ma'shar's *Introductorium maius* or Hermann's *De essentiis*. Not all these borrowings may be substantiated, for in three cases there are other, closer parallels in the writings of the late Roman philosopher Apuleius, whom Bernard is known to have read.[79] Particularly important for its astrological doctrines was the *Asclepius*, Apuleius's translation of a Greek treatise on the philosophy of the Egyptian sage Hermes Trismegistus, which survives in around twelve French manuscripts from the eleventh and twelfth centuries, and in excerpts in a further six Apuleian florilegia.[80] It was from the *Asclepius* in fact, and not from Abu Ma'shar, that Bernard drew his dynamic picture of the generation (*generatio*) of fire descending to effect the nourishment (*nutrimen*) of the lower elements.[81] The same Hermetic treatise, rather than Hermann's *De essentiis*, was likewise the inspiration behind Bernard's account of the imposition of forms on matter by the Oyarses associated with each of the seven planets and with the formative region of the fixed stars (II.5.5–18; II.3.11.1).[82]

---

79 See W. Scott and A. S. Ferguson, *Hermetica: The Ancient Greek and Latin Writings which Contain Religious or Philosophic Teachings Ascribed to Hermes Trismegistus*, 4 vols. (Oxford: Clarendon Press, 1924–1936), IV: xlvi; R. B. Woolsey, "Bernard Silvester and the Hermetic *Asclepius*", *Traditio* 6 (1948): 340–44; and McCrimmon, "Classical Philosophical Sources", 47–106, who surveys the entire Apuleian corpus, and finds only relatively insignificant traces of *De Platone et eius dogmate*, and "no unmistakable echoes of *De deo Socratis*" (85).

80 Klibansky and Regen, *Handschriften*.

81 *Cos.* I.4.2.1–4: "Ignis namque ethereus, sociabilis et maritus, gremio Telluris coniugis affusus, *generationem* rerum publicam, quam de calore suo producit ad vitam, eam inferioribus elementis comodat *nutriendam*." Compare these italicised words with those of *Asclepius* 2, in Apuleius, *De philosophia libri*, ed. Moreschini, 41.5–8: "De caelo cuncta in terram et in aquam et in aera. Ignis solum, quod sursum versus fertur, vivificum, quod deorsum, ei deserviens. At vero quicquid de alto descendit, *generans* est, quod sursum versus emanat, *nutriens*." Lemay, *Abu Ma'shar*, 264–65, was right to note a similar idea in Abu Ma'shar, but there the distinction is not between generative and nutritive qualities, but in John of Seville's translation between *agentia* and *patientia* (Abu Ma'shar, *Liber introductorii maioris ad scientiam judiciorum astrorum* II.4, ed. R. Lemay, 9 vols. (Naples: Istituto Universitario Orientale, 1995–1996), V: 74–75), and in Hermann of Carithia between *activa* and *passiva* (*Introductorium* II.4, ed. Lemay VIII: 31; Hermann of Carinthia, *De essentiis*, ed. Burnett, 90.24–25).

82 *Cos.* II.3.11.8–10: "Oyarses igitur circuli, quem Panthomorphon Grecia, Latinitas nominat Omniformem, formas rebus omnes omnibus et associat et asscribit." Lemay, *Abu Ma'shar*, 269, n. 1, found parallels in Hermann's *De essentiis*; but compare more directly *Asclepius* 19, ed. Moreschini, 58.5–8: "XXXVI, quorum vocabulum est Horoscopi, id est eodem loco semper defixorum siderum, horum οὐσιάρχης vel princeps est, quem παντόμορφον vel omniformem vocant, qui diversis speciebus diversas formas facit." The Greek Usiarch

And finally, it was not from Abu Ma'shar, but from another text in the Apuleian corpus, the pseudo-Aristotelian *De mundo*, that Bernard took the notion of the stars being composed of a fifth, unchanging element, which allowed them to be reliable guides to future events.[83]

Thus in three of Lemay's five alleged borrowings from Arabic astrology, there is a more convincing case to be made for Bernard's use of widely available Apuleian sources. In the remaining two cases, however, there is more compelling evidence that Bernard was influenced by the Arabic astrologer. One concerns the *Cosmographia*'s account of the firmament as an "elementing nature" ("elementans natura", I.4.7.8), who transforms the elemental qualities in matter into *elementa*, which can then in turn become *elementata*.[84] Whereas *elementa*

(οὐσιάρχης) may have come to Bernard in the corrupted form of 'Oyarses'. In the same passage (*Cos.* II.3.11.6–8) Bernard noted the impossibility, due to the perpetual motion of the heavens, of one form being born identical to another: "Namque impossibile est formam unamquamque alteri simillimam nasci horarum et climatum distantibus punctis." This statement comes verbatim from *Asclepius* 35, ed. Moreschini, 80.7–10: "Cum itaque haec duo, ex quibus constant formae, corpora et incorporalia sint, impossibile est formam unamquamque alteri simillimam nasci horarum et climatum distantibus punctis ..."

83 Lemay, *Abu Ma'shar*, 262, n. 4, compared the phrase "Si enim ex elementis ..." from *Cos.* II.3.10 (quoted in full below) with the italicised phrase in Abu Ma'shar, *Introductorium* I.2, ed. Lemay, VIII: 9.279–86: "Omnis vero philosophie superneque indaginis auctoritate et ratione constat substantiam stellarium corporum nec ex aliquo elementorum huius mundi effictam, nec ex pluribus vel omnibus congestam. *Si enim ex hiis elementis* esset, eam elementarie prolis necessitas consequeretur: generatio, corruptio, augmenta, detrimenta, resolutio cetereque id genus alterationes. Que cum illic aliena sint, tam circulorum celestium quam stellarium corporum substantiam ex quinta quadam alia natura consistere ratio concludit." T. Ricklin, "Die lateinische Entdeckung der Quintessenz: die *Philosophia* des Daniel von Morley", in *Metaphysics in the Twelfth Century: On the Relationship among Philosophy, Science and Theology*, ed. M. Lutz-Bachmann, A. Fiodora and A. Niederberger (Turnhout: Brepols, 2004), 85–112 (96–98), shows that the Aristotelian doctrine of a fifth element was transmitted through a variety of Latin sources. The closest to Bernard's wording is Apuleius, *De mundo* I, ed. Moreschini, 149.3–8: "caelum ipsum stellaeque caeligenae omnisque siderea compago aether vocatur, non, ut quidam putant, quod ignitus sit et incensus, sed quod cursibus rapidis semper rotetur, elementum non unum ex quattuor, quae nota sunt cunctis, sed longe aliud, numero quintum, primum ordine, genere divinum et inviolabile." Compare *Cos.* II.3.10.1–7: "Ether omnisque compago siderea non elementale est conpositum, sed ab elementis numero quintum, ordine primum, genere divinum, natura invarabile. Si enim ex elementis, natura quorum convertibilis, celum stelleque celigene substantiam contraxissent, certum nichil, nichil veridicium nuntiarent. Circulus igitur, et ambitor et extimus, nec quidem ignis est nec ignitus."

84 *Cos.* I.4.7.1–2: "Precedit Yle, Natura subsequitur elementans; elementanti Nature elementa, elementis elementata conveniunt."

are the pure elements that exist only as transcendent archetypes, *elementata* are the compounds of elemental qualities found in all physical things. Lemay believed the term *elementatum* to have been coined by John of Seville, and that its appearance in the writings of William of Conches, Bernard Silvestris and Daniel of Morley demonstrated the influence of Abu Ma'shar on these philosophers.[85] This thesis would require that William's *Philosophia*, in which the term also appears, was composed after 1133, the date of John's translation. However, William's own statements about the chronology of his career suggest that he wrote the *Philosophia* in the years between 1125 and 1130.[86] Whether William himself coined the term, or whether it derived initially from an early-twelfth-century commentary on the *Isagoge Iohannicii ad artem Galieni*, which survives in Bodl. Digby 108, remains unknown.[87] Nonetheless, there were clearly two senses to this peculiar twelfth-century term, one of which derives from a medical context, as in William of Conches, and the other from an astrological one, as in John of Seville.

William's focus in the *Philosophia* was on refining Constantine the African's definition of *elementum* as "a simple and minimal particle".[88] Following Constantine's insistence that these particles are perceptible only by the intellect, William likened them to the letters that make up a syllable.[89] The *elementa* are the qualities of cold, hot, dry and moist, and they stand in a similar relation to the visible *elementata* of fire, air, water and earth as letters do to syllables, for each *elementatum* is composed of two *elementa*, and the names

85 Lemay, *Abu Ma'shar*, 176–79 (William), 263–68 (Bernard), 323–28 (Daniel).

86 Dronke, "New Approaches", 128–31; I. Caiazzo, "The Four Elements in the Work of William of Conches", in *Guillaume de Conches*, ed. Obrist and Caiazzo, 3–66 (4–6).

87 The second theory was proposed by Elford, "William of Conches", 312, and finds strong support in Caiazzo, "Four Elements", 11–13.

88 *Philosophia* I.22, ed. Maurach, 28: "Cum ergo illae simplae et minimae particulae elementa sunt, quae est frigida et sicca, terra est; quae frigida et humida, aqua est; quae calida et humida, aer est; quae calida et sicca, ignis est. Cum igitur haec IIII quae videntur ex illis composita sunt, illud in quo dominantur particulae frigidae et siccae, nomine illius elementi dicitur terra; in quo frigidae et humidae, aqua; in quo calidae et siccae, ignis. Si ergo illis digna velimus imponere nomina, particulas praedictas dicamus 'elementa', ista IIII quae videntur 'elementata'."

89 *Philosophia* I.22, ed. Maurach, 27: "Sed quaerat aliquis: 'Ubi sunt elementa?' Nos vero dicimus: 'In compositione humani corporis et aliorum sicut littera est in compositione syllabae, etsi non per se'." This analogy was a commonplace of Epicurean atomism, and could be found in Isidore, *Etymologiae* XIII.2.4: "Unus autem atomus est, quia insecabilis. Sic et littera: nam orationem dividis in verba, verba in syllabas, syllabam in litteras. Littera, pars minima, atomus est, nec dividi potest."

of the *elementata* are simply shorthand for the particular combination of elemental qualities that dominate at any one time. John of Seville's distinction between *elementum* and *elementatum* also relies on the idea that unchanging *naturae* lie behind the varied *naturalia* of the physical world.[90] This passage occurs in Abu Ma'shar's account of the four principles of generation, which are *species*, *naturae* (i.e. *elementa*), *compositio*, and *naturalia* (i.e. *elementata*). The dense syntax of interlocking cognates (*elementa, elementans, elementata*) in John's rendering of Abu Ma'shar conveys a sense of the ontological unfolding of these generative principles. Whereas William employed the *elementa/elementata* distinction to clarify the difference between perception and reality, John's use of these overlapping lexical forms suggests that generation is a seamless process, whose origin can be traced back to a being who acts but is not acted upon. This being is the "Deus Creator benedictus et altissimus" who set the planets in motion, initiating the process of generation.[91]

Comparing these discussions with the *Cosmographia*'s account of *elementata*, it is easy to see where Bernard's affinities lie. He writes:

> Precedit Yle, Natura subsequitur elementans; elementanti Nature elementa, elementis elementata conveniunt. Sic principia principiis, sed a principe principio, coheserunt. Nisi celum, nisi motus sidereus illis quas inportat varietatibus afficiat elementa, pigra iaceant, iaceant ociosa … Est igitur elementans Natura celum stelleque signifero pervagantes, quod elementa conveniant ad ingenitas acciones. (I.4.7.1–9)

90 *Introductorium* I.4, ed. Lemay, V: 27.918–29: "Similiter etiam dicitur quod nature sint antiquiores naturali per ordinem, et iam fuerint elementa et non elementata. Et non erit directio elementati nisi per temperamentum elementorum, nec directio elementorum in elementatis nisi per temperamentum compositionis, et non erit compositio nisi per compositorem, nec elementatum nisi per causam. Et quia impossibile est ut sit compositum id quod composuit semetipsum, et sit elementatum idem quod elementavit suam essentiam; et si ita est, necesse est ut sit causa elementans elementatum ex elementis et componens compositum ac differens inter species animalium et sementum substanciarum ex elementis et speciebus." Hermann avoided the term *elementatum* and its cognates in his less literal translation and in his *De essentiis*: see Lemay, *Abu Ma'shar*, 266.

91 Abu Ma'shar, *Introductorium* I.4, ed, Lemay, V: 27.930–35: "Patet igitur nunc quod Deus Creator benedictus et altissimus posuerit planetis significationes et motus naturales, et quia quod efficitur ex virtutibus motus eorum naturalium in quatuor angulis sit compositio naturalium et differentia universarum specierum ex naturis et speciebus, per hoc quod est in speciebus ex differentiis."

> Raw matter comes first; then follows elementing nature. The elements are based on this elementing nature, and elemented substances on the elements. Thus the principles of existence depend on principles of their own, but these in turn depend on one sovereign principle. For if the firmament and the movement of the stars did not endow the elements with that capacity for change which they transmit, these would remain sluggish and inactive ... This elementing nature, then, is in fact the firmament, and those stars which traverse the circle of the Zodiac, for it is these that arouse the elements to their natural activity.

As in John of Seville, Bernard invokes a "sovereign principle" from which generation unfolds, and he likewise conveys a sense of that unfolding through grammatical variation: "elementans ... elementa ... elementata / principia ... principiis ... a principe principio". And like John's *causa elementans*, Bernard's *natura elementans* is the firmament, whose motions form the *elementa* into composite *elementata*. Taking into account the meaning, style and context of these passages, it is clear that Bernard's deployment of the *elementata* relates more closely to Abu Ma'shar's celestially oriented discussion of generation than to William's medically inspired account of the elements.[92]

The terminology of *elementata* was relatively widespread among twelfth-century philosophers.[93] But the close textual parallel discussed above offers compelling evidence that Bernard knew Abu Ma'shar's *Introductorium maius* in the translation by John of Seville. A further indicator of Bernard's knowledge of the Arabic astrologer may be found in the fifth and final of Lemay's source texts, namely Abu Ma'shar's astrological forecast of the Incarnation, which

---

92 In other discussions of the elements Bernard adopted more Conchian formulations ("elementa, partes mundi primarias partesque partium" (I.4.1.3–4), and "elementa—partes mundi primarie partesque partium" (II.9.7.7–8)), but in neither instance did he apply the *elementum/elementatum* distinction: see B. Pabst, *Atomtheorien des lateinischen Mittelalters* (Darmstadt: Wissenschaftliche Buchgesellschaft, 1994), 137–38.

93 See T. Silverstein, "*Elementatum*: Its Appearance among the Twelfth-Century Cosmogonists", *Mediaeval Studies* 16 (1954): 156–62; Gregory, *Anima mundi*, 204–6; R. McKeon, "Medicine and Philosophy in the Eleventh and Twelfth Centuries: The Problem of the Elements", *The Thomist* 24 (1961): 211–56; C. Burnett, "*Omnibus convenit Platonicis*: An Appendix to Adelard of Bath's *Quaestiones naturales*", in *From Athens to Chartres*, ed. Westra, 259–81 (268); and I. Caiazzo, *Lectures médiévales de Macrobe: les Glosae Colonienses super Macrobium* (Paris: Vrin, 2002), 60–65 (with Caiazzo, "Four Elements", 8, n. 10), who argues that William of Conches had already, before the *Philosophia*, used the term *elementatum* in his glosses on the *De consolatione Philosophiae*, a variant reading that had been rejected by that work's editor (*Glosae super Boetium*, ed. Nauta, lv–lxi).

we shall examine more closely in the following chapter. John's 1133 translation dates from around a decade before the *Cosmographia*'s publication, and it was the more popular of the two versions, surviving in around twelve twelfth-century manuscripts.[94] And even if some of his celestial doctrines derive from Apuleius, Bernard incorporated these into an astrological framework much more developed than anything to be found in that source. Revising Martianus Capella, Bernard shifted the didactic content of his allegory away from a traditional conspectus of the seven arts, to focus on a contemporary account of *scientia naturalis* enriched by medical and astrological thinking. Why then, we must now ask, did Bernard consider this kind of science to be such an important human activity?

The Purpose of Natural Science

The purpose of medical learning in the *Cosmographia* may be illuminated by Bernard's presentation of the allegorical persona of Silva. Silva derives from Calcidius, who used the term to translate Aristotle's word for matter, *hyle*, which means 'timber' as well as 'woodland', whereas *silva* denotes only the latter. Calcidius considered *silva* to be a crucial concept for understanding the *Timaeus*.[95] In his commentary, *silva* is a principle of potentiality, a material substrate invoked to explain where elemental qualities exist at the moment of their transformation from one state to another.[96] For example, when a solid (cold and dry) turns into a liquid (cold and moist), dryness turns into moisture, and at the moment of its transformation it is said to exist *in silva*.

94 Abu Ma'shar, *Introductorium*, ed. Lemay, IV: 4–5. By contrast, there is no firm evidence that Bernard knew the more recent *De essentiis* (1143) of Hermann of Carinthia. Only one manuscript of the work is known to survive from the twelfth century (*De essentiis*, ed. Burnett, 63–66), and the parallels adduced by Hermann's editor could have stemmed from common sources: thus, for Bernard's discussion of elemental generation and nutrition (*Cos.* I.4.2), see *De essentiis* 61rE, but also *Asclepius* 2; for his consideration of the circular nature of time (*Cos.* I.4.11), see *De essentiis* 70rC, but also *Asclepius* 29–30; and for his use of the phrase "ymaginariam hominis ymaginem" (*Cos.* II.11.5.15), see *De essentiis* 79rF, but also *Timaeus* 29C, ed. Waszink, 22.16.

95 See Calcidius, *Commentarius* CCLXVIII–CCCLV, ed. Waszink, 273–345, with the discussions of Dronke, *Spell of Calcidius*, 25–30; J. C. M. van Winden, *Calcidius on Matter: His Doctrines and Sources. A Chapter in the History of Platonism* (Leiden: Brill, 1959); and J. R. O'Donnell, "The Meaning of 'Silva' in the Commentary on the *Timaeus* of Plato by Chalcidius", *Mediaeval Studies* 7 (1945): 1–20.

96 Calcidius, *Commentarius* CCCXVII, ed. Waszink, 313.

In the *Cosmographia*, Bernard turned this elusive concept into a vividly depicted allegorical character. His Silva is a neglected and abused mother figure, representing the raw potential of matter before Noys informs it with ideas and shapes it into the geometrical elements of the cosmos.[97] The universe born of this process is then wedded to Endelechia, who corresponds to the *anima mundi* of the *Timaeus*, the animating spark of cosmic life. Endelechia is charged with ensuring that the unruly impulses of Silva are restrained in the heavens. "Whilst maintaining the promise of hospitality," Bernard wrote of Endelechia, "she does not share her temple nor assent to the presence within it of any violator of that promise."[98]

However, the powers of Endelechia diminish the further she moves from her celestial home (I.2.16.5–10), which means that she cannot guarantee on Earth the kind of elemental harmony that obtains in the heavens. The "renegade angels" ("angelos . . . desertores", II.7.11.2), for example, who hover above the Earth, have been "only slightly purged of the ancient evil of Silva."[99] As another inhabitant of the sublunar realm, man too faces the constant threat of Silva's elemental disruption:

> Unde homines, quia locum incolunt inquietum—tumultus instar veteris—motus permutationum necesse est experiri. Quippe que de celo sideribusque decesserat potentius expurgata, in inferioribus remansit ad plurimum—Silve necessitas influentis. (II.5.21.3–7)

> So mankind, inhabiting this unquiet region—the very image of the ancient chaos—must be subject by necessity to the motion of its turmoil. For the necessity of ever- flowing Silva, which had been rigorously purged from the heavens and the stars, remains fully active in the lower regions.

---

97 For Bernard's allegorical handling of Silva, see Stock, *Myth and Science*, 97–118; M. Asper, "Silva Parens: zur allegorischen Technik des Bernardus Silvestris", in *Fabrica: Studien zur antiken Literatur und ihrer Rezeption*, ed. T. Baier and F. Schimann (Stuttgart: Teubner, 1997), 129–47; and P. Godman, *The Silent Masters: Latin Literature and its Censors in the High Middle Ages* (Princeton, NJ: Princeton University Press, 2000), 278–79.

98 *Cos.* I.2.15.15–16: ". . . fide quidem hospitii reservata, cum expugnatore tabernaculi sui non participat nec consentit."

99 *Cos.* II.7.10.4–6: "Summa dei diligencia minus defectiusque purgati, ab antiqua Silve malicia angustissima brevique linea recesserunt." The second half of this sentence would seem to mean not that the evil spirits "have been contained within extremely narrow bonds" (*Cosmographia*, trans. Wetherbee, 108), but that they have withdrawn from the ancient evil of Silva by only a very narrow margin.

This "motion of turmoil" (*motus permutationum*) is what distinguishes the habitat of the human soul (the microcosm) from the tranquillity of the macrocosmic world soul. It impairs the smooth functioning of reason that is symbolised by Plato's account of the harmonious rotation of the Same and the Different within the soul (*Timaeus* 34C–37C). The Neoplatonists had dramatised the split between macrocosm and microcosm as a narrative of the soul's descent from the heavens into the body. This *descensus animae* was carefully described by Macrobius, whom Bernard followed closely, writing that the souls of men clustered in the firmament around the Tropic of Cancer, gloomily waiting "with faces fit for a funeral" ("vultibus quibus itur ad exequias", II.3.8.6–7) for their descent into the prisons or tombs of their corporeal forms.[100]

For man, the elemental turmoil of the sublunar world is somewhat alleviated by Physis' creation of balanced complexions in the human body (II.13.5–7). But these are not entirely effective safeguards against the physiological imbalances that threaten health. This is the best that may be said:

> Compacta est humanitas quantum curari potuit, vel indigencia vel redundancia multo parcius felicitatem operis attemptante. (II.13.8.7–9)
>
> Human nature has been as carefully wrought as possible, with lack or excess presenting scarcely any threat to the harmony of the work.

Crucial here is the qualifier "scarcely" (*parcius*). For when Physis creates man she must work with the "elemental leftovers" of creation, and these are tainted with "the stains and ineradicable evil of Silva."[101]

Some support for man's fragile physiology comes from the flora and atmosphere of the grove called Gramision, where "the first man dwelt as a guest, but too briefly for a guest."[102] But outside this paradise grove, in the wilderness of the fallen world, man is more susceptible to the unruliness of Silva. Knowledge

---

100 *Cos.* II.8.33–34: "Sed que compedibus, que carcere clausa tenetur, | En quasi corporea mole sepulta iacet!" Compare Macrobius, *Commentarii in somnium Scipionis* I.11.11–12, ed. Willis, 47. For the history of this motif, see P. Courcelle, "Tradition platonicienne et traditions chrétiennes du corps-prison", *Revue des études latines* 43 (1965): 406–43; and I. Tolomio, "'Corpus carcer' nell'alto medioevo: metamorfosi di un concetto", in *Anima e corpo nella cultura medievale*, ed. C. Casagrande and S. Vecchio (Florence: SISMEL, 1999), 3–19.

101 *Cos.* II.13.4.1: "elementorum reliquias"; II.12.19–20: "maculas et inextricabile Silve … malum."

102 *Cos.* I.3.335–36: "incoluit … | Hospes—sed brevior hospite—primus homo." See below, Chapter 3, 116–17.

of medicine, then, will be an important contribution to human well-being, for it can sustain a healthy body in which the human soul may—unimpeded—emulate the ordered movements of the world soul. Man is partly a creation of Physis, and is thereby instilled with the medical learning proper to this goddess: as Bernard puts it, Physis "presents to the various uses of mankind whatever allays disease, whatever is conducive to health, and whatever pleasurably arouses sensual desire."[103] This includes knowledge of the curative effects ("remediales ... effectus", II.9.7.10–11) of plants, herbs and stones.

This kind of information would have been readily available to Bernard in such scientific manuscripts as Tours, Bibliothèque municipale 789, a twelfth-century manuscript from the cathedral of Saint-Maurice, Tours, that contains medical and botanical extracts from Pliny (*Historia naturalis* XXV–XXVI), one of the earliest copies of the *De lapidibus* of Marbod of Rennes (*c.* 1035–1123), and Odo of Meung's late-eleventh-century *De viribus herbarum*, commonly attributed to Macer Floridus.[104] These are among the encyclopaedic sources that Bernard employed to create his versified catalogue of creation in the third chapter of "Megacosmus", which contains references to 126 species of plants, and includes a miniature herbal of fifty-six lines, summarising some of the medical applications of these plants.[105] Bernard's catalogue poem, then, provides a sample of the kind of scientific learning that he thought to have both a practical application and a theological rationale in the elemental disorder of the sublunar world.

Similar emphases may be found in Bernard's comments on the purpose of astrology. His idea of the value of this science may best be seen by examining the ways in which the *Cosmographia* echoes and revises its other principal model in the mixed form, namely Boethius's *De consolatione Philosophiae*. In the *Consolatio*, it is the aim of Philosophia to cure Boethius of his dejected state of mind by recalling him to a proper understanding of man's place in the cosmos. She wants to correct his sense of misfortune by demonstrating that the vicissitudes of *fortuna* should be understood as part of a wider pattern of

103 *Cos.* II.9.3.1–4: "Quicquid occurit morbis, quicquid conciliat sanitatem, quicquid deliciosos voluptate sensus irritat ... in diversa mortalium commoda ... subministrat."

104 Collon, *Catalogue général*, 592–95.

105 *Cos.* I.3.361–416. See N. F. Palmer, "Plant Names in the *Cosmographia* of Bernardus Silvestris", *Scientiarum Historia* 20 (1994): 39–56; and L. Moulinier, "Abbesse et agronome: Hildegarde et la botanique de son temps", in *Hildegard of Bingen: The Context of her Thought and Art*, ed. C. Burnett and P. Dronke (London: Warburg Institute, 1998), 135–56 (141–42).

cosmic order proceeding from the *providentia* of divinity. Philosophia intends to dispel the external *tenebrae* that have overwhelmed Boethius, and restore him to the inner *lux* of his own reason.[106] This cure involves a process of recollection, for, as Philosophia tells Boethius, "if the muse of Plato speaks the truth, man but recalls what he once knew and lost."[107] Boethius had been engaged in the recollection of philosophy before his imprisonment, but has since turned his back on such activities to pen the elegiac verses with which the work begins. Philosophia hopes to recall Boethius to his old ways, when, she reminisces, "you used to describe for me with your stick (*radius*) the courses of the planets, and to model human ethics and the whole conduct of life on the patterns of the celestial order."[108] To understand the order of the cosmos will be to understand man's place within it: thus to achieve a full understanding of himself and his predicament, Boethius must be, as William of Conches put it in a gloss, both an *astronomicus* and a *physicus*.[109]

This Boethian conception of science as self-understanding within the cosmos may be traced back to Pythagorean ideas in the *Timaeus*. There Plato argued that the soul of man is governed by the same *anima* that governs the cosmos, and that the harmonic ratios of the *anima mundi* must also be established in the human soul, to control the unruly impulses of man's material body.[110] This requirement accounts for the importance attached to the senses of sight and hearing in the *Timaeus*. Hearing allows man to perceive musical harmony, which, "having motions akin to the revolution of our soul,"[111] may serve to harmonise sublunar discord within the soul. Likewise, sight permits man to contemplate and "imitate" the ordered motions of the heavens, and thereby to "correct the vagaries of his own mind."[112] A version of this Platonic-

---

106 Boethius, *De consolatione Philosophiae* I, m. 2.1–5, ed. Bieler, 3: "Heu, quam praecipiti mersa profundo | Mens hebet et propria luce relicta | Tendit in externas ire tenebras | Terrenis quotiens flatibus aucta | Crescit in immensum noxia cura!"

107 *De consolatione Philosophiae* III, m. 11.15–16, ed. Bieler, 60: "Quodsi Platonis Musa personat verum, | Quod quisque discit immemor recordatur."

108 *De consolatione Philosophiae* I, pr. 4.4, ed. Bieler, 7: "cum mihi siderum vias radio describeres, cum mores nostros totiusque vitae rationem ad caelestis ordinis exempla formares."

109 William of Conches, *Glosae super Boetium* I, m. 2, ed. Nauta, 54.206–15.

110 *Timaeus* 42E–44C, ed. Waszink, 38–40.

111 *Timaeus* 47D, ed. Waszink, 45.3–4: "consanguineas habens commotiones animae nostrae circuitionibus".

112 *Timaeus* 47BC, ed. Waszink, 44.17–23 (quoted phrase in italics): "deum oculos hominibus idcirco dedisse, ut mentis providentiaeque circuitus, qui fiunt in caelo, notantes eorum similes cognatosque in usum redigerent suae mentis, circuitusque animae, qui

Pythagorean idea was also important to Stoic philosophy, which emphasised the interconnection of physics, ethics and logic, on the basis that the *logos* that directs reason and behaviour is the same *logos* that directs nature, and through possession of this *logos* man will attain a state of harmony with nature that is also a state of virtue.[113] Cicero provided a succinct account of this view, in a summary of the Stoic division of philosophy. Like *logica*, Cicero wrote, the Stoics viewed *physica* as an indispensable science, because "the starting-point for anyone who is to live in accordance with nature is the system and governance of the whole universe."[114] Variations on this theme of the ethical value of cosmology reached the Middle Ages through a number of channels, including Cicero himself, Seneca, Apuleius, Lactantius and Augustine.[115] The most influential authority on this topic, however, was Boethius, whose *Consolatio* combines the Stoic tradition with Pythagorean ideas about the numerical structure of creation, recently revived by the Greek Neoplatonist, Iamblichus.[116] In the *Consolatio*, as we have seen, it is through the mathematical contemplation of the heavens that Boethius will be reacquainted with the order of the cosmos, and thereby develop ethical defences against "fortune with its random blows" ("indiscreto ictu fortunam", II, pr. 2.12).

Aspects of this tradition are present in the *Cosmographia*, but they are adapted to Bernard's distinctive conception of the value of science. As in the *Timaeus*, man is portrayed as a microcosm created from the *anima mundi* (Bernard's Endelechia) and from the confused mass of elements left from the

animadversiones seu deliberationes vocantur, quam simillimos efficerent divinae mentis providis motibus placidis tranquillisque, perturbatos licet, confirmatoque ingeneratae rationis examine, dum imitantur aplanem mundi intellegibilis circumactionem, *suae mentis motus erraticos corrigant*."

113 M. L. Colish, *The Stoic Tradition from Antiquity to the Early Middle Ages*, 2 vols. (Leiden: Brill, 1985), I: 22–23, 36.

114 Cicero, *De finibus bonorum et malorum* III.73, ed. C. Moreschini (Munich: Teubner, 2005), 120.794–802 (quoted phrase in italics): "Physicae quoque non sine causa tributus idem est honos, *propterea quod, qui convenienter naturae victurus sit, ei proficiscendum est ab omni mundo atque ab eius procuratione*. Nec vero potest quisquam de bonis et malis vere iudicare nisi omni cognita ratione naturae et vitae etiam deorum et utrum conveniat necne natura hominis cum universa; quaeque sunt vetera praecepta sapientium, qui iubent 'tempori parere', et 'sequi deum' et 'se noscere', et 'nihil nimis', haec sine physicis quam vim habeant (et habent maximam) videre nemo potest."

115 J. Pfeiffer, *Contemplatio Caeli: Untersuchungen zum Motiv der Himmelsbetrachtung in lateinischen Texten der Antike und des Mittelalters* (Hildesheim: Weidmann, 2001), 27–104.

116 Pfeiffer, *Contemplatio Caeli*, 105–40; D. J. O'Meara, *Pythagoras Revived: Mathematics and Philosophy in Late Antiquity* (Oxford: Clarendon Press, 1989).

formation of the world.[117] Bernard too valued *contemplatio caeli*, as shown by his citation of the response of Empedocles to one who asked why he was alive: "That I might behold the stars; take away the firmament and I will be nothing."[118] Likewise Urania, in her descent through the heavens, which is an allegory of the soul's formation, affirms the value of *contemplatio caeli* when she declares:

> Mens humana michi tractus ducenda per omnes
> Ethereos, ut sit prudentior:
> Parcarum leges et ineluctabile fatum
> Fortuneque vices variabilis,
> Que sit in arbitrio res libera, quid ve necesse,
> Quid cadat ambiguis sub casibus. (II.4.31–36)

> Man's mind's to be led by me on a course through all
> The celestial sphere, to be schooled
> In the Fates' decrees, the destiny none avoid,
> And fickle Fortune's shifting lot.
> He'll learn what's freely willed, what's set in stone,
> And what derives from chance events.

At first, this appears to be simply an epitome of some of the key lessons imparted to Boethius by Philosophia in Book V of the *Consolatio*, namely the importance of understanding the relations between fate and fortune, and the distinction between free will, necessity and chance.[119] Metrically indeed, the alternating hexameters and tetrameters of this poem (the only time such a form appears in the work) would appear to imitate the metre of *Consolatio* I.3, an extended simile in which Boreas dispels the clouds that had obscured the prisoner's vision—and allows him to perceive once again the ordered grandeur of the cosmos.[120] However, as Urania continues her syllabus of celestial

---

117 See R. Finckh, *Minor Mundus Homo: Studien zur Mikrokosmos-Idee in der mittelalterlichen Literatur* (Göttingen: Vandenhoeck und Ruprecht, 1999), 116–58.

118 *Cos.* II.14.45–46: "Querenti Empedecles quid viveret, inquid: 'Ut astra | Inspiciam; celum subtrahe—nullus ero.'" Calcidius, *Commentarius* CCLXVI, ed. Waszink, 271.5–7, attributes a version of this dictum to another pre-Socratic philosopher, Anaxagoras.

119 Bernard also followed Boethius (*Cons.* I, m. 2.10–22)—and Vergil before him (*Aen.* I.740–46)—in framing his survey of natural scientific topics as a list of indirect questions.

120 See William of Conches, *Glosae super Boetium*, ed. Nauta, 66.36–38: "Sed Boreas, qui fugat illam noctem, philosophiae comparatur, quae ut Boreas purgat corda a nube, id est obscuritate doloris."

studies, there emerges a picture rather different from the Boethian model. For the mind, Urania says, will also know:

> Que virtus stellis, et quanta potentia celo
>   Et quis sidereis vigor axibus,
> Quid valeant radiis duo lumina, quinque planete. (II.4.41–44)

> What powers have the stars, and how much force the heavens,
>   What might the starry poles possess,
> What strength the sun and moon and planets five.

What Urania envisages is not just the mathematical charting of the heavens practised by Boethius with his *radius*, but as the terms *virtus*, *potentia*, *vigor* and *valeant* suggest, a study of the power and influence of celestial bodies. For Bernard, this astrological knowledge is inscribed into the firmament, which the creator used to "prefigure all that may come to pass through the law of fate."[121] And this knowledge is accessible to those who can decipher it, since, as Bernard boldly suggests, the creator had fashioned the heavens "so that ages to come might be beheld in advance, signified by starry measures."[122] Bernard, then, understood the activity of *contemplatio caeli* not as a means of reconciling oneself to the inevitability of *fortuna*, but rather as a way of arming oneself against it through astrological expertise.

Jens Pfeiffer has correctly observed that the *Cosmographia* revised the Latin tradition of *contemplatio caeli* sketched above, for Bernard implied that the deliberations of providence fell within man's understanding.[123] For this reason Bernard may also be said to have revised Macrobius, who insisted that philosophers should not employ a *narratio fabulosa* when treating the supreme deity (whom the Greeks call *tagaton*, that is 'the good'), or the mind (*noys*) emanating from that god: instead, they should use the language of similes and analogies (*similitudines et exempla*).[124] When discussing the supreme deity,

121 *Cos.* I.3.33–34: "Scribit enim celum stellis, totumque figurat | Quod de fatali lege venire potest."

122 *Cos.* I.3.57–58: "Sic opifex, ut in ante queant ventura videri | Secula, sidereis significata modis."

123 Pfeiffer, *Contemplatio Caeli*, 269: "Die Sterne sind jedoch nicht nur die Mittler in der Ausführung der Providentia, sondern zeigen dem Betrachter auch, welche künftigen Ereignisse zu erwarten sind. Sie sind damit die sichtbaren Zeichen der in der Providenz der Noys bereits vorgegeben, in der Zeit jedoch noch unerfüllten Ereignisse."

124 Macrobius, *Commentarii* I.2.14, ed. Willis, 6.22–7.4: "Ceterum cum ad summum et principem omnium deum, qui apud Graecos τἀγαθόν, qui πρῶτον αἴτιον nuncupatur, trac-

Tugaton, Bernard adhered to this prescription, deploying a range of theological formulas ("theologis ... argumentis", II.5.1.10) that belong to the apophatic tradition of theology primarily associated with pseudo-Dionysius. In a number of striking passages, Bernard writes that Tugaton inhabits "an abode of purified and uncontaminated light" ("purgati defecatique luminis ... mansionem", II.5.1.8–9), which produces a "glittering darkness" ("splendorem ... caliginem", II.5.3.5–6), and that he is "concealed, unknown, by the majesty of his divine darkness" ("Divine licet maiestate caliginis abscondatur incognitus", II.7.3.5–6). Bernard broke the Macrobian rules, however, when he came to Noys, the mind of God. For in the *Cosmographia* Noys is clearly integrated into the dramatic action of the *narratio fabulosa*, and, despite the grammatical masculinity of the Greek νοῦς, is assigned a feminine gender in common with the other allegorical personae of the work.[125] Bernard places her among the Thrones, the third rank of the celestial hierarchy.[126] The fact that Noys appears within the celestial sphere, and that she enjoys "prior knowledge of things" ("rerum cognitio prefinita", I.2.13.5), suggests that through contemplation of the heavens man too may attain knowledge of the future.

In the *Cosmographia*, then, the science of astronomy takes on decidedly astrological overtones. Meanwhile the theory of music, the sister-discipline of astronomy in the Platonic-Pythagorean tradition, is notably absent from Bernard's cosmology. There is no mention of the musical theory of the *anima mundi*, nor of its application to the construction of man.[127] And whilst Bernard

---

tatus se audet attollere, vel ad mentem, quem Graeci νοῦν appellant, originales rerum species, quae ἰδέαι dictae sunt, continentem, ex summo natam et profectam deo: cum de his inquam loquuntur summo deo et mente, nihil fabulosum penitus attingunt, sed siquid de his adsignare conantur quae non sermonem tantum modo sed cogitationem quoque humanam superant, ad similtudines et exempla confugiunt." William of Conches was also dissatisfied with the Macrobian theory of *narratio fabulosa*, but in his case with Macrobius's dismissal of the philosophical value of ancient fables recounting actions inappropriate to divinity. On this, see P. Dronke, *Fabula: Explorations into the Uses of Myth in Medieval Platonism* (Leiden: Brill, 1974), 25–30.

125 Jolivet, "Les principes féminins".

126 *Cos.* I.3.17–18: "Pura throni legio, quibus insidet ille profundus | Spiritus et sensus, mensque profunda Noys." Contrast with *Cos.* I.3.11–12: "super, immo superne, | Extramundanus creditur esse deus"—presumably a reference to the abode of Tugaton outside the cosmos.

127 Bernard does at one point refer to the "artifices numeros et musica vincla'"(I.1.22) sought by Silva, but as has been observed by M. Kurdziałek, "Der Mensch als Abbild des Kosmos", in *Der Begriff der Repraesentatio im Mittelalter: Stellvertretung, Symbol, Zeichen*, ed. A. Zimmermann (Berlin: Walter de Gruyter, 1971), 35–75 (48, n. 46): "[Bernard] entwickelte aber keine Konzeption der Welt und des Menschen, die in dieser Formulierung enthalten war."

retains Plato's emphasis on *contemplatio caeli* as the chief benefit of sight, his discussion of hearing omits any reference to music (II.14.49–86). It could be said that medicine, the subject of Physis, has replaced the study of music in Bernard's reorganisation of the hierarchy of the arts. In the cases of both astrology and medicine, scientific knowledge is designed to counteract the destabilising effects of Silva, who is not, like Boethius's Fortuna, a chimera to be dispelled by rational reflection, but an active force in the sublunar world, whose malevolence may be combated by practical measures.

Having said this, it would be wrong to suggest that Bernard had total confidence in the power of natural science to redeem fallen human existence. Despite his generally optimistic vision of the cosmos, there are also some more pessimistic tones that intrude into the *Cosmographia*, particularly in its account of man.[128] Most significantly, on each of the three instruments with which Urania, Natura and Physis are endowed to help them create man (respectively the Mirror of Providence, the Tablet of Destiny and the Book of Memory) there is reflected, etched and inscribed the hierarchy of created things, each time concluding with only a faint and barely discernable image of man. Man is the final piece in creation, but does this make him the worthy consummation of the work ("operis digna consumatione", II.3.1.4–5) or merely a cosmic afterthought, cobbled together from elemental leftovers ("elementorum reliquias", II.13.4.1)? As Winthrop Wetherbee has remarked, "the murkiness of life as viewed from the existential vantage-point of Bernardus' Nature and Physis poses a clear challenge to philosophy. It raises the question of whether unaided human understanding is capable of attaining more than a tenuous and uncertain purchase on natural reality."[129] Such reservations about natural science come more fully into focus in the two poems, based on pseudo-Quintilian's *Declamationes maiores*, that are found in the final folios of the Tours manuscript discussed at the end of the previous chapter. It is surely no coincidence that one of these, *De gemellis*, is based on a declamation with a medical theme, and the other, *Mathematicus*, on one that concerns astrology.[130] Both poems can be read as thematic sequels to the *Cosmographia*, poems that transplant the model human being created in the final sections of

128 Minkova, "*Cosmographia*", 135, concludes that "*Megacosmos* is a rather optimistic picture of the universe, and this tone is maintained in *Microcosmos*. Yet man, who is introduced in the universe with the hope of being its crown, unavoidably brings the imperfections of his nature with himself, and the optimisim in the work is steadily diluted by more pessimistic views."

129 Wetherbee, "Philosophy, Cosmology and the Twelfth-Century Renaissance", 48.

130 Of the nineteen pseudo-Quintilianic declamations, these two (IV and VIII) are the only ones concerned extensively with astrology and medicine.

that work into human society, and explore some of the problems that he will face in Bernard's cosmos.[131]

De gemellis begins with a picture of twin boys, born in ancient Rome and blessed with all nature's perfections:

> Quos sic Nature manus ingeniosa potentis
> Finxerat ex anima uel corporeis elementis,
> Ut meminisse queat nichil in rerum genitura,
> Cui sit tantus honos uel tam speciosa figura.[132]

> The nimble hand of mighty Nature did them mould
> In such a way from bod'ly elements and soul
> That in vain could she from nature call to mind
> A thing of so much grace, of such a shapely kind.

Perfection of birth, however, will not remove the twins from the fickle wheel of fortune ("rota fortune, nunquam raroue fidelis", l. 23). As youths they are both afflicted by the same mysterious disease, which undermines their complexions and the five senses, whose constitution was the subject of *Cos.* II.13–14—with the result that "Nature lies dejected, forgetful of her own goods" ("Sic oblita iacet rerum Natura suarum", l. 40). In desperation the father seeks the counsel of two physicians and, in a departure from its source, the poem specifies that one of these is Greek (l. 42). The doctors set to work, using Greek diagnostic methods that were becoming better known in the twelfth century through the availability of Latin translations of Theophilus's *De urinis* and Philaretus's *De pulsibus*.

> Sed nec in urinis uel pulsibus inspicientes
> Morborum causas potuerunt scire latentes. (ll. 45–46)

> The disease's hidden causes they could not discern
> By inspecting urine and the pulses each in turn.

131 The *Mathematicus* was certainly completed before 1153, the latest possible date for its earliest known manuscript, BN lat. 5129 (*Mathematicus*, ed. Prelog *et al.*, 12, 15); that the work postdated the *Cosmographia* is suggested by internal evidence discussed below, n. 138. The date of *De gemellis* is unknown.

132 *De gemellis*, ll. 7–10, ed. Edwards, "Poetic Invention", 198. I have capitalised Edwards' "nature" in l. 7: with her *manus*, and as the subject of *queat* (l. 9), Nature is surely here a personification.

Even with the latest medical expertise, then, the physicians fail to fulfill the promise made by Noys in the *Cosmographia* that man "will see clearly the principles shrouded in darkness, so that Natura may keep nothing hidden."[133] It is only at the cost of sacrificing one of the sons for dissection that the doctors discover the disease's cause and therefore its cure. The remainder of the poem is concerned with the lawsuit brought by the wife against her husband for approving a course of action that has robbed her of a son. Medicine may (or may not: see ll. 72–76) have saved one of the twins, but in any case even the most advanced medical practice was not able to prevent tragedy.

Bernard's *Mathematicus* is a longer and much more complex poem, and this is not the place to discuss the many problems of interpretation that it presents. One of its themes, however, is once again the human costs incurred by the limited nature of scientific knowledge. The poem begins with a noble couple in ancient Rome who have failed to conceive a child. The wife seeks the advice of an astrologer, who reassures her that she will have a son, and indeed that he will become king of Rome—but he will also kill his father.[134] The son is born and, like the twins in *De gemellis*, he is "of such god-like appearance that he could scarcely be believed of human substance."[135] The mother cannot bear to kill her son and so, feigning his death, sends him away to be raised by another, where he is given the name Patricida as a reminder of his destiny. As a youth he excels in the seven arts (l. 136), which in Bernard's numbering include physics and medicine in place of grammar and geometry. The young man's knowledge of astronomy extends to an astrological understanding of "the orbits of the constellations and the human destinies they rule", the same elevating knowledge that was promised to man by Urania in *Cosmographia* II.4.[136] Ironically though, and in spite of his name, Patricida remains unaware of his destiny. Fearing idleness and its attendant vices, he enters the military and distinguishes himself, and eventually, after averting imminent defeat at the

133 *Cos.* II.10.35–36: "Viderit in lucem mersas caligine causas, | Ut Natura nichil occuluisse queat."

134 The astrologer is strictly speaking correct when he says that his prophecy contains "nil ... ambigui" (l. 60), but due to the twelve-word separation of *patrem* from its modifying adjective *suum* (rhetorical hyperbaton), the full horror of what he foretells is only apparent with the final word of the prophecy (ll. 55–58): "Inter felicis fatalia commoda uite | Romulee dominus et pater urbis erit. | Patrem—sed taceo nisi quod premis, arguis, instas—| Occidet fato sic agitante suum." I quote from the edition and translation of Deidre Stone, with punctuation added.

135 *Math.* 87–88: "In nascente fuit tante deitatis imago | Uix potuit credi materialis homo."

136 *Math.* 127: "Astrorumque uices humanaque fata sub astris." See above, 81–82. Similar parallels with *Cos.* II.8 are drawn by Ratkowitsch, "Astrologie und Selbstmord", 194, n. 42.

hands of Hannibal's Carthaginians outside the walls of Rome, he is made ruler of the city.

At this point the mother reveals to her husband the identity of the new monarch, and the parents visit the royal palace to warn their son of his terrible destiny. Upon meeting him, the mother praises Patricida in the following terms:

> Nostraque iustificas priscumque reducis in aurum
> Secula nec ferri nomen habere sinis.
> Nature prelarga manus te pretulit orbi
> Et dedit in solo munere cuncta semel.[137]

> You vindicate our time and lead us back to long-gone gold,
> Nor do you let this age assume the name of iron.
> The bounteous hand of Nature lifted you above the world,
> And granted by this single gift all things at once.

Patricida's virtues are so great that he alone seems to reverse the decline from a golden age that had been inscribed onto Natura's Tablet of Destiny. His stature is comparable to that of Pope Eugenius III, whom in the *Cosmographia* Bernard praised in the same terms as "a single gift, by which the deity gives all at once."[138]

After his parents leave him, Patricida delivers a moving soliloquy in which he resolves to avoid his destiny, asking:

> Nostra quid ethereis mens est cognatior astris
> Si dure Lachesis triste necesse ferat? (ll. 639–40)

> How kindred is our mind to the celestial stars if it
> Must bear the grim demands of stubborn Lachesis?

---

137 *Math.* 503–6. In verse 503 I prefer the reading in Prelog *et al.* (with the implied object *tempora* drawn from the previous line) over Stone's unmetrical "Naturamque iustificas priscumque reducis aurum."

138 For the fall from a golden age to an age of iron, see *Cos.* II.11.9.8–9: "Secularis illa continentia, purioribus ex auro iniciata principiis, paulatim degenerante materia, in ferrum visa est terminare"; and for God's gift of Eugenius, *Cos.* I.3.55–56 (italics added): "Munificens deitas Eugenium comodat orbi, | Donat et *in solo munere cuncta semel.*" This would suggest a prior date of composition for the *Cosmographia*, since it seems likely that the intertext is designed to emphasise the excellence of Patricida (and thus the pathos of his fall) by association with a good pope, rather than to associate Eugenius—before whom the *Cosmographia* was recited—with a tragic pagan.

In the *Cosmographia*, Noys ordained that "the heavenly powers, the stars, the firmament, will speak to man, and Lachesis reveal to him her plans."[139] Lachesis, of the three Fates the one responsible for measuring out the thread of human life, may reveal to man her intentions; but here the *Mathematicus* insists that, however great he may be, man has no power to change his life's allotted span. Patricida appears before the senate to request a favour whose nature he cannot reveal. The senators gladly agree, but when they learn that the king's wish is to commit suicide, they rescind. The poem ends abruptly with Patricida divesting himself of his royal robes, and thus asserting his freedom as a private citizen to defy the senate's ruling.[140] The reader is left uncertain of whether it will be Patricida or his father who will die—or even both, as in the fourth declamation, which imagines the son accidentally killing the father even as he takes his own life.[141] In any case, astrology has not prevented tragedy. At best, as in *De gemellis*, scientific knowledge may have reduced the human cost by allowing Patricida a voluntary act of self-sacrifice; at worst, it may have cost two lives rather than one. Both poems, then, use extreme but vividly imagined scenarios to underline the limitations of the scientific knowledge promised by the *Cosmographia*.

In this chapter, to summarise, we have explored how Bernard Silvestris echoed and revised some of the major curriculum *auctores* of the twelfth-century schools to create a compelling and contemporary vision of natural science. His treatise in the mixed form shares with Martianus Capella's a promise of philosophy's restorative influence on the human soul, but it departs from Martianus's schematisation of the seven arts by promoting a natural philosophy that is not simply reducible—as it had been before in the medieval West—to the mathematics of the quadrivium. Much of this science was indebted to Plato, Calcidius and the range of late antique Latin Platonists who were well known in Bernard's milieu. But by recasting Plato's creative *dii deorum* as the sister-goddesses Urania, Natura and Physis, Bernard placed special emphasis on the medicine (Physis) and astrology (Urania) of the new Arabic-Latin trans-

139 *Cos.* II.10.33–34: "Dii superi stelleque sibi celumque loquetur, | Consilium Lachesi notificante suum."

140 See Dronke, *Fabula*, 138, who writes that "[t]he human mind, for Bernard as for his hero, has a sovereign freedom that fate and the stars cannot crush ... Physical necessity may be ineluctable, and the mind itself so conditioned by the world around that it seems in an impasse; but, for Bernard as for Boethius, the mind, if unencumbered enough by its physical habitation, need not be affected."

141 *Declamationes*, ed. Håkanson, (IV.23), 84.7–17.

lations. I have attempted to bring the special characteristics of this science into focus by reading the *Cosmographia* in conjunction with its other ancient model in the mixed form, Boethius's *De consolatione philosophiae*. For whilst Bernard echoed Boethius on the value of contemplating the heavens, his emphases on medicine and astrology insist that this is not purely a contemplative exercise, but a practical one that might mitigate the problems of inhabiting a world elementally disordered by the ever-presence of Silva. The *Cosmographia*, then, is a call to study, an exhortation to rediscover the natural philosophy that man's soul absorbed in its descent through the cosmos. By doing so, man may create within himself an image—a microcosm—of the world soul that guarantees the tranquillity of the heavens. Nonetheless, Bernard's ideal natural philosopher—theorised in the *Cosmographia*, and brought to life in the *Mathematicus* and *De gemellis*—remains a potentially tragic figure, for despite his best efforts he will never be exempt from the vicissitudes of sublunar life.

CHAPTER 3

The Theology of the *Cosmographia*

If the *Cosmographia* was written at a time of great creativity in natural philosophy, it was also the product of a period of intense religious and theological activity in France. Bernard's life coincided with a renewed movement for monastic reform, involving a return to the traditions of the early church and the establishment of the new Cistercian and Carthusian orders. Meanwhile, in the northern French schools, theologians were bringing a new rigour to the definition of Christian orthodoxy by writing works that offered systematic accounts of Christian doctrine, instead of the piecemeal treatment of earlier scriptural commentaries. This process would culminate, a decade after the *Cosmographia*, in the production of the most important theological textbook of the Middle Ages, the *Sentences* of the Parisian theologian Peter Lombard.[1] The conjunction of these developments with the ambitious philosophy of the schools rekindled debates from the days of the early church about the value of secular learning in Christian education. Such debates focused primarily on the use of the newly fashionable linguistic arts in the discussion of the Eucharist and the Trinity, but also, in the cases of Peter Abelard and William of Conches, they concerned the legitimacy of using natural science to expound the creation story of Genesis. These controversies led to the papal condemnation of Abelard's teaching at the Council of Sens (1141) and to the letter condemning William of Conches's *Philosophia* by the Cistercian monk William of Saint-Thierry (*c.* 1140–1142).

Early critics were inclined to see the *Cosmographia* as a similarly heterodox product of the French schools in the 1140s. Bernard's first editors remarked on his "sceptical" attitude towards theology,[2] and others spoke more forcefully of a "complete abandonment of the orthodox views",[3] and tendencies towards the heresies of dualism and pantheism.[4] The Benedictine authors of the

1 For general overviews, see G. R. Evans, *Old Arts and New Theology: The Beginnings of Theology as an Academic Discipline* (Oxford: Clarendon Press, 1980); J. Leclercq, "The Renewal of Theology", in *Renaissance and Renewal*, ed. Benson and Constable, 68–87; M.-D. Chenu, "Masters of the Theological 'Science' ", in his *Nature, Man and Society in the Twelfth Century*, 270–309; Southern, *Scholastic Humanism*, I: 102–33; and Colish, *Peter Lombard*, I: 42–90.

2 *De mundi universitate*, ed. Barach and Wrobel, xv.

3 T. O. Wedel, *The Medieval Attitude toward Astrology* (New Haven, CT: Yale University Press, 1920), 34.

4 De Wulf, *Histoire de la philosophie médiévale*, 233.

 | DOI 10.1163/9789004268357_005

Histoire littéraire de la France were perhaps the first moderns to criticise Bernard's theology, but because they believed him to have been Bernard of Chartres—the venerable master who led his students in pious Christian colloquia—they had to conjure into existence a professor whose written work somehow lacked the good sense he showed in the classroom.[5] And yet for all these verdicts, there is no evidence that Bernard was prosecuted or his work suppressed. The *Cosmographia* survives in over fifty manuscripts, and none of those examined for this study appears to have been expurgated to remove theologically dubious material.[6] Furthermore, a well-known gloss in one of these manuscripts informs us that the *Cosmographia* was recited in France before Pope Eugenius III, who did not apparently supress the work, as he did the teachings of Gilbert of Poitiers during the same French sojourn (1147–1148).[7]

5 *Histoire littéraire de la France*, XII: 264: "Mais on doit se ressouvenir qu'en tout art, et surtout dans le grand art d'écrire, il y a loin de la conoissance des règles à l'application, du talent de montrer à celui d'exécuter. Il ne fut pas donné à notre habile professeur de franchir absolument cet intervalle, et de mettre autant d'élégance et de correction dans ses écrits, que de méthode et de lumière dans ses leçons." For the daily prayers in Bernard of Chartres' classes, see John of Salisbury, *Metalogicon* I.24, ed. Hall and Keats-Rohan, 53.69–76.

6 By contrast, a twelfth-century copy of Bernard's *Mathematicus* (Bern, Burgerbibliothek 710, ff. 82r–92v) removes all references to the pagan gods, astrology and Carthage. The copyist treated the manuscript's texts of Vergil, Horace and Ovid in a similar way. See *Mathematicus*, ed. Stone, 213, 216.

7 Bodl. Laud misc. 515, f. 188v (glossing *Cos.* I.3.55–56, which mentions the pope): "Iste Eugenius fuit papa in cuius presencia liber iste fuit recitatus in Gallia, et captat eius benivolenciam." The final phrase of this gloss has invariably been translated as "it [i.e. the book] won his benevolent approval" (*Cos.*, 2) or "gained his favour" (*Cosmographia*, trans. Wetherbee, 135, n. 85). It actually means "it *seeks to obtain* his favour/approval", the present tense *captat* referring not to Eugenius's response but to the intention of the flattering couplet that it glosses: "Munificens deitas Eugenium comodat orbi, | Donat et in solo munere cuncta semel". This is important because Peter Godman, adopting the normal mistranslation of the Bodleian gloss (*Silent Masters*, 233), has argued that Eugenius "expressed an approval that was a factor in the swift and lasting success of the *Cosmographia*. Its author had satisfied the rules of a procedure Abelard failed to follow, and was granted the twelfth century's version of a *nil obstat*" (335). Godman's claim that there was a "procedure" in place for the papal pre-censorship of texts relies in turn on a misreading of Peter Abelard's account of the criticism he faced at the Council of Soissons in 1121. The passages reads (*Historia calamitatum*, ed. J. Monfrin (Paris: Vrin, 1959), 87): "Dicebant [emuli mei] enim ad dampnationem libelli satis hoc esse debere quod nec romani pontificis nec Ecclesie auctoritate eum commendatum legere publice presumpseram." Abelard is not saying here that his enemies thought that failure to submit his treatise for papal approval *was* sufficient for it to be condemned, but that it *ought to be* ("Dicebant... satis hoc esse *debere*"). This has been pointed out several times by Constant

Later critics have tended to handle this evidence in one of two ways. Étienne Gilson, writing in 1928, argued that Bernard "wishes to stay in line with the teachings of theology".[8] Gilson pointed to a number of biblical allusions in the *Cosmographia* and to certain structural parallels with the Book of Genesis. These led him to claim that the *Cosmographia* was really a kind of hexameral commentary, which in the tradition of the Greek and Latin Fathers employed pagan cosmology to elucidate the creation story of Genesis. For Gilson, the apparent paganism of the *Cosmographia* was merely a consequence of Bernard's admiration for the ancient poets, which led him "to dress up Christian substance in a disconcertingly pagan form".[9] Gilson's Christianising interpretation did not win many supporters.[10] Ernst Robert Curtius, writing twenty years later, conceded the presence of biblical material in the *Cosmographia* but thought it peripheral, arguing that "Bernard represents a pagan humanism which eliminates everything Christian except for a few ultimate essentials." According to Curtius, Bernard's thinking was not theological but rather informed by ancient literary evocations of the goddess Natura, which inspired a revival of the pagan fertility cult in the *Cosmographia*.[11] The emphasis on classical, pagan sources in the *Cosmographia*, and the marginalising of biblical

---

Mews, for example in his "St Anselm and Roscelin: Some New Texts and their Implications, I: The *De incarnatione verbi* and the *Disputatio inter Christianum et gentilem*", *AHDLMA* 66 (1991): 55–98 (65–66). Mews's reading solves the problem noticed by G. B. Flahiff, "Ecclesiastical Censorship of Books in the Twelfth Century", *Mediaeval Studies* 4 (1942): 1–22 (4), that Abelard's testimony appears to be the only twelfth-century evidence of what Godman calls a "procedure" of papal pre-censorship: it is, in fact, nothing of the sort. The 1147/48 recitation of the *Cosmographia* need not, then, be viewed as a doctrinal inquest. It may instead have been a learned court entertainment—perhaps at the church of Saint-Martin, of which Eugenius was titular abbot. For the contemporary recitation of lengthy works of Latin literature, see Baudri of Bourgueil's ecphrasis of the historical and cosmological motifs decorating the bed-chamber of Countess Adela of Blois, in a poem which concludes with the promise that its bearer will recite it to the countess: "Misi qui nostrum reddat recitetque libellum" (*Adelae Comitissae*, l. 1367, in *Poèmes*, ed. and trans. Tilliette, 43).

8 Gilson, "La cosmogonie", 12: "... veut rester d'accord avec les enseignements de la théologie".

9 "La cosmogonie", 24: "... à revêtir de manière déconcertante un fonds chrétien d'une forme païenne".

10 Silverstein, "Fabulous Cosmogony", provides a review of reactions to Gilson's article and a more nuanced development of his Christianising thesis.

11 E. R. Curtius, *Europäische Literatur und lateinisches Mittelalter* (Bern: Francke, 1948), 119: "Ihm tritt in Bernhard ein heidnischer Humanismus gegenüber, der das Christliche bis auf ein Existenzminimum ausscheidet"; trans. Trask, *European Literature*, 112.

and theological ones, has continued in more recent criticism. Brian Stock, for instance, devoted only a few lines of his study to the biblical allusions, commenting that "they are few in number, indicating that Bernard was simply fonder of the classics".[12] Barbara Newman has likewise downplayed their importance, writing that the *Cosmographia* "includes a few biblical allusions, but chiefly as historical asides," and that they "seem little more than decorative".[13]

Diametrically opposed though the Christian and paganist readings are, they both involve a distinction between supposedly essential features of the *Cosmographia* and those that are merely peripheral. For Gilson, "the allegories borrowed from the ancient poets" made up a literary veneer that could be peeled away to reveal the core Christianity of the work.[14] Bernard lacked that special unity of form and content that Gilson found in a truly great writer such as Dante.[15] Meanwhile, for Curtius and his successors, the classicising form and pagan sources of the *Cosmographia* made it an essentially non-Christian composition, on which biblical allusions were scattered either as a minimum doctrinal requirement (Curtius) or merely as rhetorical ornament (Newman). Neither of these positions seems entirely satisfactory. For one thing, the *Cosmographia* is a very different kind of endeavour from those didactic works that employed classical metres for their alluring stylistic qualities, whilst disowning the corrupting morality of the ancients: we have already encountered such an approach in the late-twelfth-century *Hexameron* of Andrew Sunesen.[16] And secondly, just because the *Cosmographia* is more deeply permeated by the philosophical sources analysed in the previous chapter, this does not mean that the smaller number of biblical and theological motifs are extraneous to the work.

The argument of this chapter will be that the theology of the *Cosmographia* should, like its science, be understood in terms of twelfth-century approaches to the interpretation of the ancient *auctores*. Bernard aspired to the status of

12 *Myth and Science*, 131.

13 *God and the Goddesses*, 64–65.

14 "La Cosmogonie", 9: "les allégories empruntées aux poètes antiques".

15 See É. Gilson, *Dante et la philosophie* (Paris: Vrin, 1939), writing of "les grands écrivains" (ix): "leurs idées font partie de leur art, et c'est leur grandeur même que ce qu'ils disent reste après eux inséparable de la manière dont ils l'ont dit." Compare Gilson, *La philosophie au moyen âge: des origines patristiques à la fin du xiv*[e] *siècle*, 2nd ed. (Paris: Payot, 1944), on the *Cosmographia* (273): "Certains passages de cette oeuvre ne sont pas sans beauté, mais il est vraiment excessif de rappeler à leur occasion, ainsi qu'on l'a fait, le grand nom de Dante et le souvenir de la *Divine comédie*."

16 See above, Chapter 1, 41.

these illustrious predecessors, and I believe it is as one of their number that he wished his *Cosmographia* to be read. What follows will suggest that Bernard conceived of his *Cosmographia* as a contribution to a long and fascinating intellectual tradition that sought to identify continuities between Greek philosophy and Judaeo-Christian theology. A product of the apologetics of the early Church Fathers, this project was revived in the twelfth century as the pagan *auctores* rediscovered by the Carolingians became more fully integrated into the scholarship of the Christian West. Twelfth-century masters employed a sophisticated lexicon of terms, such as *integumentum* (covering) and *involucrum* (wrapping), to describe how the richly nuanced language of the pagans concealed certain prefigurations of what would become 'revealed' religion.[17] Such attempts raised questions about the extent to which the kind of allegorical interpretation common in biblical exegesis should be permitted in secular texts. Could the same approach that turned the male lover of the Song of Songs into a figure of Christ also reveal the world soul of Plato's *Timaeus* to be an allegory of the Holy Spirit? These were concerns that came to a head in the decade that Bernard was writing, in those polemics against the teachings of Peter Abelard and William of Conches mentioned at the start of this chapter.

The following discussion begins with a survey of Latin patristic thought about the theological value of pagan thought, and follows the development of such thinking into the twelfth century. My argument will be that parts of the *Cosmographia* should be read as a reaction to what some saw as the encroachment of natural philosophy into theology in the teachings of Abelard and William of Conches. Bernard's reaction consisted of a return to, and an imitation of the polysemous language of the pagan *auctores*, a language in which hidden truths of philosophy and theology awaited detection by a sufficiently learned reader. I shall support this argument by exploring three of the biblical and theological themes that are woven into the fable of Bernard's work. The first, the Trinity, relates to long-established debates about the relationship between Platonic cosmology and the biblical creation narrative. The other two topics, the Fall and the Incarnation, offered Bernard opportunities to explore tensions between theology and the new medical and astrological science of the twelfth century. I shall conclude by suggesting that Bernard's 'integumental' theology—his recreation of the numinous allegories of antiquity—should not, as some have suggested, be understood as an exercise in the evasion of

17 For discussion of these critical terms, see É. Jeauneau, "L'usage de la notion d'integumentum à travers les gloses de Guillaume de Conches", *AHDLMA* 24 (1975): 35–100; Dronke, *Fabula*, 13–78; Stock, *Myth and Science*, 49–54; Wetherbee, *Platonism and Poetry*, 36–48; and Bezner, *Vela Veritatis*.

censorship through figurative language. Rather, I propose that Bernard designed his obscurities deliberately to lead his readers into a deeper consideration of the mysteries of creation.

Platonism and Christianity

Twelfth-century scholars inherited from the Church Fathers two distinct ways of describing the compatibility of Platonic and Mosaic cosmology. The first was a historical argument, which claimed that Plato had travelled to Egypt to listen to the prophet Jeremiah or to read the Scriptures, from where he derived certain cosmological doctrines in his *Timaeus*. Widely supported in Greek patristics, this idea occurs in the Latin theologians Ambrose (*Expositio Psalmi CXVIII*, XVIII.4) and Jerome (*Ep.* 53), from whom it found its way into the brief biographies of Plato included in the *accessus* to copies of the *Timaeus* dating from the eleventh century.[18] The chronological priority of the Old Testament prophets as the original cosmologists also found support in Cassiodorus's claim, following the first-century *Jewish Antiquities* of Josephus, that Abraham had transmitted arithmetic and astronomy to the Egyptians.[19] In antiquity Egypt was widely thought to have been the birthplace of philosophy. Martianus Capella, for instance, described how Philology, before her apotheosis, vomited up a stream of writings concerning the seven arts, which were then inscribed on stone and placed in a cave in Egypt.[20] Cassiodorus's contention was no doubt intended to suggest that the Christian appropriation of pagan philosophy—the "spoiling of the Egyptians" (Exodus 3:21–22; 12:35–36)—should be understood as a return to its owners of what was rightfully theirs.[21] The Old

18 A. S. Riginos, *Platonica: The Anecdotes Concerning the Life and Writings of Plato* (Leiden: Brill, 1976), 65, 91; Somfai, "Eleventh-Century Shift", 11–12.

19 Cassiodorus, *Institutiones* II.3.22, ed. R. A. B. Mynors (Oxford: Clarendon Press, 1937), 131.21–132.4: "Scire autem debemus Ioseppum Hebreaorum doctissimum in primo libro *Antiquitatum*, titulo nono, dicere arithmeticam et astronomiam Abraham primum Aegyptiis tradidisse, unde semina suscipientes, ut sunt homines acerrimi ingenii, excoluisse sibi reliquas latius disciplinas"; followed by Isidore, *Etymologiae* III.25.1.

20 Martianus Capella, *De nuptiis* I.35–37, ed. Willis, 42; and see *Cos.* I.3.49: "Egyptus parturit artes."

21 According to a gloss on Martianus in BN, lat. 8674, f. 41r, the sons of Noah inscribed the contents of the arts onto two pillars before the flood. The first pillar, made of brick, was destroyed; the second, made of stone, survived in Egypt, and it allowed Abraham to teach astronomy when he was in exile there. See C. E. Lutz, "Remigius' Ideas on the Origin of the Seven Liberal Arts", *Medievalia et Humanistica* 10 (1956): 32–49. The source for this idea

Testament prophets were in fact the original philosophers, and from this idea it was a short step to the interpretation of Scripture as philosophy. Thus Jerome (following Origen) wrote that physics, ethics and logic were treated respectively by the three Solomonic books of Ecclesiastes, Proverbs and the Song of Songs—with Genesis also included under physics, and the Gospels under logic.[22] This was an inspired polemical manoeuvre. For Genesis could now be seen as the archetypal authority on natural science, and one whose legitimate heir was not Plato and the Greek philosophical tradition, but Solomon and the *contemptus mundi* of Ecclesiastes.

Some theologians, however, were sceptical about this historical explanation of the resemblances between Platonic and Mosaic cosmology. Augustine at first accepted Ambrose's claim that Plato had studied with Jeremiah in Egypt (*De doctrina christiana* II.43), but later, in the seminal treatment of Platonism and Christianity in Book VIII of *De civitate Dei*, he rejected the idea after a reconsideration of the chronology led him to conclude that Plato had lived around a century after Jeremiah, and that he had died sixty years before the publication of the Greek Septuagint (*De civitate Dei* VIII.11). Augustine nonetheless maintained that there were striking parallels between certain teachings of Scripture and the ethical, logical and scientific ideas of some pagan philosophers, especially the Platonists. These, he ventured, were to be accounted for not in terms of the direct influence of the prophets on the Greeks, but rather by the proximity to truth that the reasoning mind may achieve without the aid of revelation.

This alternative to the historical explanation posited the existence of what Augustine called *theologia naturalis*. This was one branch of a widely employed tripartite scheme, the other two branches being 'civil' and 'fabulous' theology, which Christians used to classify and evaluate the ways in which pagans treated divinity.[23] In *De civitate Dei* Augustine assessed in detail the value of pagan

would appear to be Josephus, *Jewish Antiquities* I.64, where Jobal, the inventor of music, inscribed two columns (one of brick, against fire, and one of stone, against water) to preserve his teachings.

22 Jerome, *Epistulae*, XXX.1, ed. I. Hilberg, 4 vols. (Vienna: Österreichische Akademie der Wissenschaften, 1996), I: 243.5–10: "quomodo philosophi solerent disputationes suas in physicam et ethicam logicamque partiri, ita et eloquia divina aut de natura disputare, ut in Genesi et Ecclesiaste, aut de moribus, ut in Proverbiis et in omnibus sparsim libris, aut de logica, pro qua nostri θεογικήυ sibi vindicant, ut in Cantico canticorum et evangeliis." Isidore, *Etymologiae* II.24.8, repeated this idea in his division of philosophy.

23 For the sources and development of the tripartite scheme, see J. Pépin, *Mythe et allégorie: les origines grecques et les contestations judéo-chrétiennes*, 2nd ed. (Paris: Études Augustiniennes, 1976), 276–392; and for Augustine's understanding of the history of phi-

theology in these three areas, and he was most sympathetic towards the claims of *theologia naturalis*, which consisted of philosophical discussions of the nature of the gods and the world. Here the pagans came closest to the truth, by virtue of the rational speculation alluded to in Romans 1:20, which says that God's "invisible realities were made visible to the intelligence through his created works, as were his eternal power and intelligence".[24] In this realm of speculation, Augustine noted, "there are none who come nearer to us than the Platonists."[25] Their conceptions of God as the author of the universe, the source of truth and the bestower of happiness were very close to Christianity. By considering the nature of the intellect, they transcended the materialism of the pre-Socratics to arrive at the truth that the creator of the immaterial mind must himself be immaterial, and they were also correct in their assertions that God was uncreated and immutable (*De civitate Dei* VIII.5–6). Finally, like hexamerists before and after him, Augustine found in the opening two verses of Genesis a strong parallel with Plato's account of the ordering of the elements (*De civitate Dei* VIII.11). All this, of course, had providential meaning.[26]

Following Augustine's dismissal of the notion that Plato had been directly indebted to the Old Testament prophets, the historical explanation of the correspondence between Platonic and Mosaic cosmology was not popular in the twelfth century.[27] Peter Abelard, for example, in the three successive revisions of his *Theologia* (*c.* 1119–*c.* 1139), quoted with approval Augustine's sceptical remarks on Plato and Jeremiah.[28] Following Augustine, Abelard viewed the

---

losophy, see G. Piaia, *Vestigia philosophorum: il Medioevo e la storiografia filosofica* (Rimini: Maggioli, 1983), 21–30.

24 *De civitate Dei* VIII.6, ed. B. Dombart and A. Kalb, 2 vols. 5th ed. (Stuttgart: Teubner, 1981), I: 331.9–13: "Ita quod notum est Dei, manifestavit eis ipse, cum ab eis invisibilia eius per ea, quae facta sunt, intellecta conspecta sunt; sempiterna quoque virtus eius et divinitas; a quo etiam visibilia et temporalia cuncta creata sunt."

25 *De civitate Dei* VIII.5, ed. Dombart and Kalb, I: 327.1–2: "Nulli nobis quam isti propius accesserunt."

26 *De civitate Dei* VIII. 5–6, 11. See also *Confessiones* VII.9 on the correspondences between Scripture and the *libri platonici*.

27 Alexander Neckam (1157–1217) still maintained this thesis (see *Cantate domino canticum novum*, cited by Ferruolo, *Origins*, 358, n. 101, from Bodl. Wood empt. 13, f. 89r), and following in the footsteps of Jerome's and Isidore's association of books of the Bible with branches of philosophy (see n. 22, above) Neckam wrote a treatise *De naturis rerum* that was designed to serve as a preface to a planned commentary on Ecclesiastes.

28 *Theologia 'summi boni'* I.57, in *Petri Abaelardi Opera Theologica, III*, ed. E. M. Buytaert and C. J. Mews, CCCM 13 (Turnhout: Brepols, 1987), 106–7; *Theologia 'scholarium'* I.181–83, in *Opera Theologica, III*, ed. Buytaert and Mews, 394–96; *Theologia christiania* I.118, in *Petri Abaelardi Opera Theologica, II*, ed. E. M. Buytaert, CCCM 12 (Turnhout: Brepols, 1969), 122.

testimonia philosophorum as an expression of natural theology, in which it was sometimes possible to discern prefigurations of Christian doctrine; thus Abelard too quoted St Paul on the *invisibilia Dei* and the *virtus* and *divinitas* of God that had been made manifest to the intelligence (Rom. 1:20). But the twelfth-century philosopher went further than Augustine by equating these three terms with, respectively, the Spirit, the Son, and the Father, and by arguing that Paul had thought the pagans to have had some knowledge of the Trinity.[29] Abelard's assessments in his *Theologia* of the *testimonia philosophorum* in relation to revelation are the most elaborate exercises of their kind in the twelfth century, and the range of doctrines for which he found pagan *integumenta* is extensive.[30] His principal concern though was to identify prefigurations of specifically trinitarian doctrine across a range of philosophical authorities (Plato, Vergil, Macrobius, the Sibylline oracles, the *Asclepius*), and the Trinity was also the topic of greatest interest to other contemporary scholars.

Speculation in this area was stimulated by the difficult question of how to understand the role of the Trinity in creation. Since Augustine (*De civitate Dei* XI.32), it had been common to read the first two verses of Genesis as implying the joint involvement of all three persons in the act of creation. The Father was the *deus* who created heaven and earth, the Holy Spirit the *spiritus* that moved over the waters, and the involvement of the Son, the "Word made flesh" (John 1:14), was revealed by the opening verse of John's Gospel: "In principio erat verbum et verbum erat apud deum." The problem lay in understanding the precise characteristics of the Trinity's involvement in creation, given the doctrinal difficulties ever attendant on discussions of this greatest of Christian mysteries. To some though, the *testimonia philosophorum* promised to fill the gaps left by Scripture, and to provide clues towards understanding the creativity of the Trinity.

---

29 *Theologia christiana* II.13, in *Opera Theologica, II*, ed. Buytaert, 138–39: "Ex quibus aperte Apostolus docet eis quoque mysterium Trinitatis fuisse reuelatum. Quod enim dictum est *inuisibilia Dei* eis fuisse reuelata, ad personam Spiritus non incongrue applicatur, qui hoc ipso quod 'Spiritus' dicitur, ex ipso nomine 'inuisibilis' insinuatur. Et quia ipse *donum Dei* uocatur, et secundum septiformem gratiam multiplex et septem spiritus quasi septem dona nonnumquam dicitur, non incongrue dictum est pluraliter *inuisibilia Dei*, hoc est Spiritum eius septiformis gratiae donorum distributorum. Verbum autem Patris Christum Dominum, Dei uirtutem et Dei sapientiam Apostolus clamat. Vnde et hoc loco cum dicitur: *Sempiterna quoque Dei uirtus*, persona Filii exprimitur. Vbi et cum subditur *et diuinitas*, quia, ut supra meminimus, maxime nomine 'dei' uel 'domini' potestatis maiestas exprimitur, persona quoque Patris expressa est."

30 T. Gregory, "Abélard et Platon", in *Peter Abelard*, ed. E. M. Buytaert (Louvain: University Press, 1974), 38–64.

Two strands of thought may be identified in twelfth-century discussions of the creative Trinity. One was directly associated with the cosmology of the *Timaeus*, the other with the emanative cosmology of Macrobius. To begin with Plato: according to a Middle Platonist reading from the second century AD, *Timaeus* 29E taught that there were three aspects of divinity evident in the created world: goodness, wisdom and power.[31] This was one of the teachings of Platonism that Augustine thought congruent with Christian theology, and in the twelfth century the goodness, wisdom and power of the creator came to be more specifically identified with the three trinitarian persons. John of Salisbury, for instance, remarked in his *Policraticus* (*c.* 1159):

> Opinio tamen illa conualuit quia in libris Platonis inueniuntur multa dictis consona prophetarum. Nam in Thimeo, dum causas mundi subtilius inuestigat, manifeste uidetur exprimere Trinitatem quae Deus est, efficientem causam constituens in potentia Dei, in sapientia formalem, finalem in bonitate quae sola induxit eum ut omnem creaturam bonitatis suae participem faceret, prout natura cuiusque capax beatitudinis esse potest.[32]

> In the books of Plato I believe we find many things consonant with the words of the prophets. In the *Timaeus*, for example, during his subtle investigation of the causes of the world, he seems clearly to express the Trinity that is God, when he locates the efficient cause in the power of God, the formal cause in his wisdom, and the final cause in that goodness which alone induced him to make all creatures sharers in his goodness, according as the nature of each one is capable of beatitude.

This equation of the Platonic triad with the Christian Trinity had some theological credibility, in the sense that *potentia, sapientia* and *bonitas* could relate respectively to the Almighty Father, the wisdom of the Son (according to the Pauline definition of Christ as "Dei sapientia" (1 Cor. 1:24)) and the grace of the

---

31 J. Whitaker, "Goodness, Wisdom, Power: A Middle Platonic Triad", in *Sophies maietores*, ed. Goulet-Cazé *et al.*, 179–94.

32 *Policraticus* VII.5, ed. C. C. J. Webb, 2 vols. (Oxford: Clarendon Press, 1909), II: 108.14–22. John may have learned these ideas from Abelard: see P. von Moos, *Geschichte als Topik: das rhetorische Exemplum von der Antike zur Neuzeit und die historiae im "Policraticus" Johanns von Salisbury* (Hildesheim: Oms, 1998), 447–48. For John's rejection of the thesis of Jeremiah's influence on Plato, see *Policraticus* I.4.

Holy Spirit.[33] Abelard appears to have been the first to make this connection explicit, and it proved to be one of the most controversial aspects of his theology.[34] The problem was that this could be viewed as an annexation of particular qualities to individual persons of the Trinity, which might then entail the existence of three separate gods. This was the understanding of the crowd who tried to stone Abelard on his arrival at Soissons for his first trial in 1121: they had, according to Abelard, "been led to believe that I was preaching and had written to prove the existence of three gods."[35] Two of the fifteen Abelardian propositions later condemned at Sens in 1141 also related to this argument.[36] And in William of Saint-Thierry's letter on the errors of William of Conches, the Cistercian monk accused the author of the *Philosophia* of repeating the trinitarian heresies of Abelard.[37]

The other strand of thought current in the twelfth century characterised the creativity of the Trinity in terms of Macrobius's account of the emanation of being.[38] According to this approach, the emanation of being from *deus*, through *mens* and *anima*, could be understood as an account of how the Father acted through the Son and the Spirit to create the world. Of particular interest to scholars in the first half of the twelfth century was the doctrine of the *anima*

33 For the further connection between the Trinity and Aristotle's four causes (known through Boethius), with *silva* as the material cause, see J.-M. Parent, *La doctrine de la création dans l'École de Chartres: étude et textes* (Paris: Vrin, 1938), 69–81.

34 See Colish, *Peter Lombard*, I: 96–131.

35 Peter Abelard, *Historia calamitatum*, ed. Monfrin, 83: "Antequam autem illuc pervenirem, duo illi predicti emuli nostri [Alberic of Rheims and Lotulf of Lombardy] ita me in clero et populo diffamaverunt, ut pene me populos paucosque qui advenerant ex discipulis nostris prima die nostri adventus lapidarent, dicentes me tres deos predicare et scripsisse, sicut ipsis persuasum fuerat."

36 Mews, "The Lists of Heresies Imputed to Peter Abelard", *Revue bénédictine* 95 (1988): 73–110 (108–10): cap. 1: "Quod Pater sit plena potentia, Filius quaedam potentia, Spiritus Sanctus nulla potentia"; cap. 14: "Quod ad Patrem, quia ab alio non est, proprie vel specialiter attineat omnipotentia, non etiam sapientia et benignitas."

37 William of Saint-Thierry, *De erroribus Guilelmi de Conchis* 2, in *Recueil d'études sur saint Bernard et ses écrits*, ed. J. Leclercq, 4 vols. (Rome: Storia e letteratura, 1962–1987), IV: 357–70 (361.104–10): "Ubi omitto interim gravem etiam Sabellianae haeresis notam, quam ipse sibi in caput suum contrahit cum de Deo dicturus Patre et Filio et Spiritu sancto Deum pro eis potentem, sapientem volentemque constituit; quod Deum qui in semetipso et ex semetipso est, quod est, ad creaturam dicit esse totum, quod est; ad ea quae proposuit veniamus." For William's argument see *Philosophia* I.3, ed. Maurach, 20–22; and for a comparison of the doctrines of Abelard and William on this subject, Gregory, *Anima mundi*, 106–21.

38 See *Commentarii* I.14.

mundi—formulated by Plato and discussed by Macrobius—for it appeared to provide an animating principle to explain how transcendent divinity operated through the Spirit in the creation of the world.[39] Abelard, for example, argued that the Holy Spirit was the world soul of the philosophers, animating the world in the same way that the human soul animates the body; and this interpretation tied in with his definition of the Spirit as *bonitas*, a vehicle for God's grace in the world.[40] Thierry of Chartres, in his commentary *ad physicam* on the Book of Genesis, proposed that the *anima mundi* was the *spiritus* that moved on the water (Gen. 1:2), and he assembled a collection of Solomonic and pagan source-texts (including the *Asclepius*, Plato and Vergil) to support this interpretation.[41] William of Conches defined the *anima mundi* as a "natural force" and, in a gloss on the Timaean hymn of *De consolatione Philosophiae* III, m.9, he equated this power with the Holy Spirit, alluding to a possible biblical authority in the form of Paul's words to the Athenians at the Areopagus, that the Christian understanding of God as He in whom "we live and move and have our being" was close to what some of the pagan *poetae* had said about divinity (Acts 17:28).[42]

---

39 Gregory, *Anima mundi*, 123–74; I. Caiazzo, "Le discussione sull'*Anima mundi* nel secolo xii", *Studi filosofici* 16 (1993): 27–63; A. Tarabochia Canavero, *Esegesi biblica e cosmologia: note sull'interpretazione patristica e medioevale di Genesi 1,2* (Milan: Vita e penserio, 1981).

40 *Theologia 'scholarium'* I.168, in *Opera Theologica, III*, ed. Buytaert and Mews, 388: "Conferant humanae animae corporibus nostris animalem uitam; conferat anima mundi, quam, ni fallor, spiritum sanctum intellexerunt philosophi, ipsis animabus nostris uitam spiritalem, distributione scilicet suorum donorum, ut sint singulae animae uitae corporum, spiritus autem sanctus uita animarum, quas uegetando ad profectum bonorum operum promouet."

41 *De sex dierum operibus* 26–27, in *Commentaries on Boethius*, ed. Häring, 566–67: "Et Mercurius in eo libro qui inscribitur *Trismegistus* hanc virtutem spiritum vocat his verbis: Fuit deus et yle quam Grece credimus mundum. Et mundo comitabatur spritus vel mundo inerat spiritus [*cf. Asclepius* 14] ... Plato vero in *Timeo* eundem spiritum mundi animam vocat [*cf. Timaeus* 34BD]. Virgilius vero de illo spiritu ita dicit: 'Principio maria ac terras celumque profundum | Lucentemque globum lune tytaniaque astra | Spiritus intus alit' [*Aen.* VI.724–26]. Hebrei vero ita de spiritu operatore locuntur. Moyses quidem ita: *Et spiritus domini ferebatur super aquas* [Gen. 1:2]. David vero sic: *Verbo domini celi firmati sunt* et cetera [Ps. 32:6]. Salomon quoque de spiritu sic dicit: *Spiritus domini replevit orbem terrarum* [Ws. 1:7]. Christiani vero illud idem Spiritum sanctum appellant."

42 *Glosae super Boetium* III, m.9, ed. Nauta, 169.522–28: "Anima mundi est naturalis vigor ... Sed ut michi videtur ille naturalis vigor est spiritus sanctus, id est divina et benigna concordia, quia divino amore et concordia habent omnia esse, moveri, vivere, crescere, sentire, discernere." Thierry of Chartres saw in the Athenian altar to the "unknown god" (Acts 17:23) a pagan interest in negative theology, as exemplified by the *Asclepius*, that

Abelard, Thierry and William were not the first Christians to have explored the notion of an equivalence between the Holy Spirit and the *anima mundi*. The identification had been widespread in Greek patristics, and Augustine had supported it in his *De immortalitate animae*, but later admitted that he had been in error.[43] The essential problem was that Plato had considered the world soul to be a creation of the Demiurge (*Timaeus* 34C–35A), and this idea of a created being did not sit comfortably with the insistence in the Nicene Creed that the Spirit was not created but rather that it "proceeds" (*procedit*) from the Father. Abelard claimed that Plato had expressed himself at this point through figurative language (*involucra*), and that, properly understood, his words were consonant with trinitarian orthodoxy.[44] But the ecclesiastical authorities were not persuaded, and the proposition "Quod Spiritus Sanctus sit anima mundi" was one of the Abelardian *capitula* condemned at Sens in 1141.[45]

The *Cosmographia* was written after the climax of these controversies in the early 1140s. It represents a step back from the direct use of Plato and other pagan philosophers to elucidate the mystery of the Trinity. For Bernard's cosmology took the form of a pagan fable similar to those of the *auctores* in whom northern French scholars of the previous two decades had been looking for prefigurations of Christianity. Just like its principal models and sources—the *Timaeus*, the *Commentarii in somnium Scipionis* and the *Asclepius*—the *Cosmographia* contains passages that suggest a correspondence between Platonic and biblical cosmology. But Bernard did not force these connections

---

foreshadowed the work of Christian theologians such as pseudo-Dionysius, who was thought to have been one of Paul's Athenian converts (Acts 17:34). See Thierry of Chartres, *Glosae super Boethii librum de trinitate* IV.8–15, in *Commentaries on Boethius*, ed. Häring, 286–87, especially IV.11, 286.74–77: "Et hoc totum testator Mercurius in eo libro qui inscribitur *Trismegistrus* [*sic*; *cf. Asclepius* 20]. Hinc est etiam quod gentiles phylosophi aram constituerant *ignoto deo* hunc attendentes de quo quis sit sciri non posset. Sed nichil predicari potest vel subici nisi unitum vocabulo."

43 Augustine, *Retractiones* I.11.4, ed. A. Mutzenbecher (Turnhout: Brepols, 1984), 35.52–58: "Sed animal esse istum mundum sicut Plato sensit, aliique philosophi quamplurimi, nec ratione certa indagare potui nec divinarum Scriptuarum auctoritate persuaderi posse cognovi. Unde tale aliquid a me dictum quod accipi possit, etiam in libro *de immortalitate animae* temere dictum notavi non quia hoc falsum esse confirmo, sed quia nec verum esse comprehendo, quod sit animal mundus."

44 *Theologia christiana* I.70, in *Opera Theologica, II*, ed. Buytaert, 101: "Dicit itaque hanc ante mundi constitutionem factam siue genitam a Deo esse; in quo Spiritus Sancti perpetuam processionem qua ex Deo Patre est, nobis, ni fallor, insinuat."

45 Mews, "Lists of Heresies", 108, cap. 3, and see also cap. 2: "Quod Spiritus Sanctus non sit de substantia Patris aut Filii."

on the reader. Instead, in the wake of a reaction against the literalist equation of Platonism and Scripture, he returned to the integumental language of the *auctores* themselves, pointing once again to certain similarities, but leaving the assessment of these similarities to the readers of his work.

The Trinity

One of the areas where the *Cosmographia* displays an integumental approach to theology is in its treatment of creation itself. For running in counterpoint to the main fable are several biblical motifs drawn from the Book of Genesis. As Gilson observed, the depiction of Silva in the work's opening lines corresponds to the void and formlessness of the earth in Gen. 1:2; the creation of the elements in *Cos.* I.2.9–11 relates to the separation of water, air and earth in Gen. 1:6–10; the creation of the cosmos from the fixed stars down to the animals and plants of Earth, narrated in the catalogue poem, mirrors the creative activity of God on the third, fourth and fifth days (Gen. 1:11–25); and the creation of man, described in the "Microcosmus", is the work of the sixth day (Gen. 1:26–31).[46] One might add that the final lines of the entire work, a couplet on the creation of man's hands and feet, allude to the traditional conclusion of a hexameral commentary, such as Ambrose of Milan's, which ends with a catalogue of man's body parts, from the head and the eyes through to the genitals and the feet.[47] Within this pseudo-biblical framework, Gilson could point to only one close verbal parallel with Genesis: the echo of Gen. 1:31 in Bernard's statement that Noys "saw that all she had made was good, and would be pleasing to God".[48] Crucially though, and unrecognised by Gilson, this line also echoes a

46 Gilson, "La cosmogonie", 7–9.

47 The contrast here is with the universe, which Bernard (following *Timaeus* 33D) had just said requires no limbs (*Cos.* II.14.179–80). In manuscripts this final section of the *Cosmographia* is frequently glossed with headings such as "de corde", "de splene", etc, which would have been reminiscent of glosses that accompany the closing anatomical section of Ambrose's *Hexameron* in a manuscript such as the twelfth-century Saint-Martin copy: Tours, Bibliothèque municipale 270, ff. 47r–49v. This parallel is one piece of evidence against the claim made by Jaeger, *Envy of Angels*, 281, that "the *Cosmographia* is unfinished". Another is the imaginative suggestion by Finckh, *Minor Mundus Homo*, 149, that the final words of the *Cosmographia*, "omnificasque manus", relate cyclically to the opening lines on Silva, suggesting the continuity of creativity through human craftsmanship.

48 *Cos.* I.2.11.2: "bona vidit omnia que fecisset, deique aspectibus placitura"; *cf.* Gen. 1:31: "Viditque deus cuncta quae fecit erant valde bona." See Gilson, "La cosmogonie", 9, n. 2.

phrase from one of Bernard's favourite sources, the Hermetic *Asclepius*, which speaks of "all that would be pleasing to the sight of god".[49] This is precisely the kind of natural theological 'coincidence' savoured by Thierry of Chartres.[50] The difference is that Bernard left this connection on the level of allusion and implication, rather than making it explicit.

The same situation obtains with another area in which Gilson argued for the essential Christianity of the *Cosmographia*, namely the doctrine of the Trinity. Here Gilson pointed to three related pieces of evidence.[51] The first was Bernard's reference to the three rays "of unified and identical brilliance" that shine from the realm of Tugaton.[52] Before this "threefold majesty" ("trine maiestati", II.5.4.1), Urania and Natura pray for the success of their venture. For Gilson, these rays were a symbol of the Trinity, and he further argued that the persons of the Trinity were represented in the fabulous narrative of the *Cosmographia* by Tugaton (the Father), Noys (the Son) and Providentia (the Spirit). Finally, Gilson claimed that Bernard was refering to the trinitarian process behind creation when he alluded to the *potentia, sapientia* and *voluntas* of the creator:

> Quisnam ergo mundo et eternitati eius audeat derogare, ad cuius continentiam causas eternas videat convenisse?—dei quidem de voluntate consensum, de sapientia consilium, de omnipotentia causas, pariter et effectum. (I.4.6.3–6)
>
> Who then would dare to disparage the universe and its eternal basis, when he may behold eternal causes working together to effect its moderation? From the will of God issues harmonious volition, from his wisdom deliberation, from his omnipotence both cause and effect.

As later scholars have observed, Gilson's identification of fabulous analogues to the persons of the Trinity was based on a misunderstanding of the text.[53]

49 *Asclepius* 3, ed. Moreschini, 42.7–8: "cuncta dei visibilis placitura". The earlier editors of the *Cosmographia*, Barach and Wrobel, read "visibus [not *aspectibus*] placitura", a variant that would bring Bernard's words even closer to the *Asclepius*.

50 See above, n. 42.

51 "La cosmogonie", 11–12.

52 *Cos.* II.5.3.1–3, 6–10: "E sedibus quidem quas Tugaton suprema divinitas habitatrix insistit, splendor emicat radiatus, non utique perfunctorius, sed infinibilis et eternus … Ex splendore igitur vel infinibili vel eterno alter se radius exerebat, ut ex primo secundoque suboriretur et tercius. Qui quidem radii, uniformes et claritatis parilitate consimiles, cum omnia collustrassent, se rursus sui fontis liquoribus admiscebant."

53 Curtius, *European Literature*, 113, n. 21; Silverstein, "Fabulous Cosmogony", 107–8.

Noys is feminine in the *Cosmographia*, despite the grammatical masculinity of the Greek νοῦς, so she cannot be the masculine Son of God; and Providentia is not a distinct character but simply another name for Noys. Endelechia, an established synonym for the *anima mundi*, would have been a more suitable candidate for the Holy Spirit but, as Gilson realised, this was ruled out by the fact that Bernard likened Endelechia to "a sphere of vast size yet *fixed* dimensions".[54]

Gilson was himself aware of some of the problems with his interpretation, but he chose to attribute these inconsistencies to Bernard's use of a poetic form:

> La pensée de Bernard apparaît donc assez flottante, sinon contradictoire, sur cette importante question; son ambition d'exposer des doctrines techniques sans renoncer à une forme littéraire peut être d'ailleurs responsable de l'indécision au moins apparente de sa pensée sur un certain nombre de points.[55]

> It seems that Bernard's thinking about this important question was rather loose, if not contradictory. Moreover, his aspiration to explain technical doctrines without abandoning a literary form may be to blame for an apparent indecisiveness in his thinking about a number of topics.

Gilson was right, up to a point. But if my reading is correct, Bernard's indecisiveness was an entirely deliberate consequence of his literary form, not an unfortunate side effect.

Thus, as with the interlacing of hexameral motifs in an overtly Neoplatonic fable, so with the Trinity there is a juxtaposition of theological and philosophical traditions that are never fully synthesised by the author. First there is the reference to the three rays shining from the abode of Tugaton, which are described in apophatic language befitting the transcendence and ineffability of the Trinity.[56] Then there are allusions to the two natural theological lines of thought that I have called the Platonic and the Macrobian. Both suggest the trinitarian nature of the creator, the first through reference to the divine attributes of *potentia, sapientia* and *voluntas*, the second by analogy with the

54 *Cos.* I.2.14.1–2: "Comparuit igitur exporrecte magnitudinis globus, terminate quidem continentie."

55 "La cosmogonie", 17.

56 *Cos.* II.5.3.3–6: "Ea igitur lux inaccessibilis intendentis reverberat oculos, aciem preconfundit, ut, quia lumen se defendit a lumine, splendorem ex se videas caliginem peperisse."

emanative triad of *deus, mens* and *anima*. Bernard perhaps wished to disassociate himself from the heterodox propositions in this field condemned in the early 1140s, for he commented that "truly there is no division in divinity", and followed this with the Pauline statement (*cf.* Phil. 4:7) that the Seraphim "repose in the peace of God, which is beyond understanding".[57] Bernard's method was rather to move fluidly between fable, hexameron, and apophatic theology, in order to suggest—rather than to insist upon—correspondences between Christianity and the *testimonia philosophorum*. He left it to his readers to resolve these various positions.

This method of integumental composition in relation to the creative activity of the Trinity has been remarked on in some of the more recent scholarship by critics dissatisfied with the simplistic choice between a 'Christian' or a 'pagan' interpretation offered by Gilson and Curtius.[58] However, Bernard's integumental method extended not only to traditional questions about the role of the Trinity in creation, but also to the new medicine and astrology that had recently begun to enrich the study of *physica* in the northern French schools, and whose significance for the *Cosmographia* was demonstrated in the previous chapter. For this new medicine and astrology brought with them doctrinal complexities of their own, such as the relationship between medical and theological anthropology on the one hand, and free will and astral determinism on the other. We shall turn first to Bernard's theological handling of the new medical science.

The Fall

The ninth chapter of the "Microcosmus" describes the secluded paradise grove in which Physis is joined by her sisters, Urania and Natura, before they embark together on the creation of man. The name of this grove has long puzzled scholars. In five of the seven manuscripts used in Dronke's edition it is called Granusion, but in the other two, Gramision.[59] The nineteenth-century editors knew only of the more common reading, which has traditionally been explained as a corruption of the Greek *hygran usian* (liquid being), a term used by Calcidius for the lower stratum of the atmosphere, in which it is possible for

57 *Cos.* II.7.5.7–8: "Non enim verum est quippiam in divinitate dividuum"; II.7.5.13: "in pace dei, que omnem sensum superat, conquiescunt."

58 Economou, *Goddess Natura*, 58–72; Whitman, *Allegory*, 218–60; Ratkowitsch, *Cosmographia*, 15–38; Pfeiffer, *Contemplatio Caeli*, 259–82.

59 *Cos.*, 159.

man to breathe.[60] Dronke, though, preferred the less frequent Gramision on the solid basis of Bernard's own explicit statement: "Nomen loco *Gramision*, quia *graminum* diversitatibus perpetuo conpubescit" ("The place is named Gramision, since it is forever bringing to maturity all kinds of plants", II.9.2.4–5); and he suggested that the name might be intended to evoke a grassy Elysium (*graminum elision*).[61]

In a paper published eight years later, Jean Jolivet agreed with this reading but thought that the coinage derived from the Latin word for lap or womb (*gremium*) combined with the Greek diminutive suffix *–ion*, so that Gramision signified a small womb.[62] Jolivet supported his derivation by analysing the language of this passage alongside that of another, from the catalogue poem in the first book of the *Cosmographia*, which is clearly a depiction of the same place.[63] He pointed to the shared language of sexuality and fertility employed in these two descriptions of the grove: Bernard wrote that it was to be found in "the flowering lap of the Earth" ("telluris gremio floridiore locus", I.3.318), in a "hidden valley" ("sinus abditus", II.9.3.3–4), where the temperate climate "impregnates the soil" ("pregnat humum", I.3.322). Furthermore, Jolivet observed that a large majority of the plants listed in the two descriptions of the grove had medical uses in gynaecology or as aphrodisiacs. The amomum and cinnamon, which are said to flourish in Gramision (II.9.4.6–7), as well as the mace, cassia, spikenard, saffron, galingale, ginger and thyme, which are listed in the catalogue poem (I.3.325–30), all appear with one or other of these

---

60 Calcidius, *Commentarius* 129, ed. Waszink, 172.3–7: "Summum enim esse locum ait ignis sereni, huic proximum aethereum, cuius corpus esse ignem aeque, sed aliquanto crassiorem quam est altior ille caelestis, dehinc aeris, post humectae substantiae, quam Graeci hygran usian appellant, quae humecta substantia aer est crassior, ut sit aer iste quem homines spirant, imus uero atque ultimus locus terrae." For this explanation, see Gilson, "La cosmogonie", 19, n. 1; followed by C. S. Lewis, *The Discarded Image* (Cambridge: Cambridge University Press, 1964), 59–60; Stock, *Myth and Science*, 187–93; and *Cosmographia*, trans. Wetherbee, 161–62, n. 64.

61 *Cos.*, 171–72.

62 Jolivet, "Les principes féminins", 301–5. Jolivet pointed out that Dronke's Elysium etymology suggested a celestial paradise for the dead rather than a terrestrial one for the living: compare *Cos.* II.5.20.9–12: "Ea quidem in parte celi, quia natura est invariabili mulcebris, et quieta, solers Grecia campos consentit elisios, et felices animas alma sacrata et nunquam desitura lucis amenitate vestiri."

63 Among several similarities, Gramision is located at the eastern limit of the Earth ("Gramision locus est Eoum ad cardinem …", II.9.2.1), just as the first grove is near to dawn and the abode of the east wind ("At potius iacet Aurore vicinus et Euro …", I.3.317); and both places are to be found in "the flowering lap of the Earth": compare "Telluris gremio floridiore locus" (I.3.318) with "florentis terre gremio" (II.9.1.12).

uses in twelfth-century southern Italian medical texts. Finally, he added, the moist and warm climate that is said to obtain in Gramision was precisely the elemental condition recommended by gynaecologists as the optimal state for conception.[64] All this fertility imagery, Jolivet argued, related to the function of Gramision as the birthplace of man, and it provided further evidence in support of the paganist interpretation proposed by Curtius of the *Cosmographia* as a work "bathed in the atmosphere of a fertility cult, in which religion and sexuality mix".[65]

Dronke has since conceded the superiority of Jolivet's interpretation, but this fertility interpretation has not met with universal approval.[66] Whilst Jolivet was correct to highlight the medico-scientific characteristics of Gramision, the flora that he discusses do not necessarily relate to gynaecology or aphrodisiacs. As the previous chapter has demonstrated, Bernard was not at the forefront of the French reception of southern Italian medicine. It would be unwise then to suppose, without more conclusive evidence, that he knew the very recent medical texts that Jolivet cites as his sources.[67] And even if these plants belonged to a common stock of gynaecological lore known to

64 See, for example, the *Flos medicinae scholae Salerni*, ll. 50–52, in *Collectio Salernitana*, ed. S. de Renzi, 5 vols. (Naples: Filiatre-Sebezio, 1852–1859), V: 3; and the *De secretis mulierum* 7–10, ed. de Renzi, IV: 5–7. Jolivet did not provide detailed references for the medicinal plants. The relevant texts are: *De aegritudinum curatione* 134–49, ed. de Renzi, II: 332–47; *De secretis mulierum* 16–34, ed. de Renzi, IV: 11–18; *Flos medicinae scholae Salerni*, ll. 950–51, 964, 996, 1167, 3408, ed. de Renzi, V: 27–28, 33, 100.

65 Jolivet, "Les principes féminins", 305; and see Curtius, *Europäische Literatur*, 119–20: "Das Ganze ist von der Atmosphäre eines Fruchtbarkeitskultus durchdrungen, in dem sich Religion und Geschlechtlichkeit vermischen"; trans. Trask, *European Literature*, 112. Jolivet repeated this claim in an essay on "Poésie et philosophie au xiie siècle", *Perspectives médiévales* 17 (1991): 51–70 (repr. in Jolivet, *Perspectives médiévales et arabes* (Paris: Vrin, 2006), 11–28 (22)).

66 P. Dronke, "Eriugena's Earthly Paradise", in *Begriff und Metapher: Sprachform des Denkens bei Eriugena*, ed. W. Beierwaltes (Heidelberg: Winter, 1990), 213–29 (repr. in Dronke, *Sources of Inspiration: Studies in Literary Transformations, 400–1500* (Rome: Storia e letteratura, 1997), 37–59 (56–57)). See though the reservations of Ratkowitsch, *Cosmographia*, 100–101, n. 192, who argues that Jolivet has over-interpreted the common trope of earth-as-womb, concluding that: "mehr als literarisches Spiel darf man in dieser Formulierung nicht sehen."

67 In fact the *De aegritudinum curatione*, for example, was not even composed until the second half of the twelfth century: see *The "Trotula": A Medieval Compendium of Women's Medicine*, ed. and trans. M. H. Green (Philadelphia: University of Pennsylvania Press, 2001), 49. Palmer, "Plant Names", 52, n. 4, is sceptical about the gynaecological reading of Bernard's plants.

Bernard via other sources, the botany and topography of Gramision do not necessarily point to the kind of paganist interpretation that Jolivet has proposed. For once again, as with the passages on the Trinity, Bernard's description includes Christian doctrines and topoi that hint at the identity of Gramision as the Garden of Eden.

The most obvious of these hints is Bernard's allusion to the Fall of man, in lines which refer to the grove as that place where "the first man dwelt as a guest, but too briefly for a guest".[68] As we have seen, Bernard adopted a Neoplatonic account of the 'fall' of the human soul from its celestial home into the prison or tomb of its corporeal form.[69] Because this position was shared by a number of Greek theologians (principally Philo of Alexandria and Origen) who argued that Eden did not have a physical location but symbolised the uncorrupted state of the soul before its 'fall' into corporeity, some have argued that Bernard too subscribed to a spiritualist conception of Eden.[70] However, despite his evident enthusiasm for Neoplatonic doctrine, it is clear that Bernard held to the orthodox position that Eden had a physical location on Earth. For Bernard's grove is embedded in a list of plants and eleven exemplary woods that contain them (I.3.339–54), and existing in space, it also exists in time: man was there more briefly ("brevior", l. 336) than a guest.[71] It would, then, be more accurate to say that Bernard entertained the idea of a double fall, incorporating both the notion of a Neoplatonic *descensus animae* and the Judaeo-Christian doctrine of physical expulsion from a terrestrial paradise.

---

68 *Cos.* I.3.335–36: "Hos, reor, incoluit riguos pictosque recessus | Hospes—sed brevior hospite—primus homo."

69 See above, Chapter 2, 77.

70 For this argument, see W. Otten, *From Paradise to Paradigm: A Study of Twelfth-Century Humanism* (Leiden: Brill, 2004), 22; and W. von den Steinen, "Les sujets d'inspiration chez les poètes latins du xiie siècle", *Cahiers de civilisation médiévale* 9 (1966): 165–75, 363–83 (375). For the Greek theologians, see A. Scafi, *Mapping Paradise: A History of Heaven on Earth* (Cambridge: Cambridge University Press, 2006), 36–43; and R. H. Grimm, *Paradisus coelestis, Paradisus terrestris: zur Auslegungsgeschichte des Paradieses im Abendland bis um 1200* (Munich: Finck, 1977), 22–71. The spiritualist conception of Eden was largely rejected in Latin theology, although it was endorsed by Eriugena (*Periphyseon* IV, 814A–848C, ed. É. Jeauneau, CCCM 164 (Turnhout: Brepols, 2000), 103–51) and by Honorius Augustodunensis, who wrote a compendium of Eriugenean thought some time between 1115 and 1130 (*Clavis physicae* 271–94, ed. P. Lucentini (Rome: Storia e letteratura, 1974), 218–40).

71 This may allude to contemporary speculation that Adam and Eve spent only seven hours in paradise. See Honorius Augustodunensis, *Imago mundi* III.1, ed. V. I. J. Flint, *AHDLMA* 49 (1982): 7–151 (123–24).

In addition, both descriptions of the creation grove are clearly indebted to a tradition of depicting Eden as a *locus amoenus*, in terms derived ultimately from the pastoral verse of Vergil, Horace and Ovid.[72] Gramision is a secluded garden of delights, situated in the East, and blessed with a perpetually temperate climate—all characteristics that appear frequently in earlier accounts of Eden.[73] In other respects, Bernard's borrowings are more idiosyncratic, and indicative both of his intimate familiarity with the Latin *auctores* and of the resonance of his own poetic imagery. Thus in the couplets that precede the lines on the Fall, Bernard writes of the stream that flows through the creation grove:

> Inter felices silvas sinuosus oberrat
> Inflexo tociens tramite rivus aque,
> Arboribusque strepens et conflictata lapillis
> Labitur in pronum murmure limpha fugax.
> Hos, reor, incoluit riguos pictosque recessus
> Hospes—sed brevior hospite—primus homo. (I.3.331–36)

> Amidst the fruitful woodland wends a winding brook,
> Often shifting its crooked course;
> Rumbling through tree-roots and jostled by stones, the liquid
> Flows fleetingly, murmuring on.
> In this brightly tinted fertile haunt, I think, dwelt man
> At first, too briefly to be called a guest.

In the first of these lines, the movement from spondees to dactyls evokes the upright trees through which the languid stream winds its way. The alliterative

72 For the classic outline of these conventions, see Curtius, *European Literature*, 183–202; and for a sketch of their fusion with theology in depictions of Eden, J. Delumeau, *History of Paradise: The Garden of Eden in Myth and Tradition*, trans. M. O'Connell (New York: Continuum, 1995), 10–15.

73 See, for example, the extensive treatments of Eden in the early Christian poets Avitus of Vienne, *De spritualis historiae gestis* I.193–237, in *The Fall of Man: De spiritualis historiae gestis libri i–iii*, ed. D. J. Nodes (Toronto: PIMS, 1985), 24–27; Claudius Marius Victorius, *Alethia* I.223–75, in *Commodius; Claudius Marius Victorius*, ed. J. Martin, CCSL 128 (Turnhout: Brepols, 1960), 137–39; and Dracontius, *De laudibus Dei* I.178–328, in *Fl. Merobardis Reliquae Blosii Aemilii Dracontii Carmina; Eugenii Toletani Episcopi Carmina et Epistulae*, ed. F. Vollmer (Berlin: MGH, 1905), 30–40. For two detailed and widely known examples from early medieval cosmography, see Isidore of Seville, *Etymologiae* XIV.3.2–4; and Rabanus Maurus, *De universo* III (PL 111, 334A–335A).

and onomatopoeic "lapillis | labitur" and "murmure" of ll. 333–34 mimic the pleasant rippling and rumbling of the water as it courses through the stones. Even before the final couplet, though, this perfect *locus amoenus* seems to foreshadow the Fall. For Bernard calls the water fleeting—"limpha fugax", an image borrowed from Horace, *Odes* II.3.12, where it stands for the transience of man's life.[74]

Bernard's model for his plant catalogue may ultimately be traced to Ovid, *Metamorphoses* X.86–106, where a host of trees come to shade Orpheus as he grieves for Eurydice.[75] Of the cinnamon, cassia, spikenard and amomum that Bernard selected as his Edenic flora, Jolivet could associate all but the final one with contemporary gynaecological theory. There is, however, a more perfect fit with another passage from the *Metamorphoses* (XV.391–400), where these aromatic plants are used by the phoenix to line her sacrificial nest. This combination of plants was in turn adopted by Lactantius, whose poem *De ave phoenice* was read as an allegory of the Resurrection, and whose selection of plants passed into the descriptions of Eden to be found in the biblical epics of Victorius and Dracontius.[76] These plants do not occur in the tradition of prose exegesis of Eden (for example, in those hexameral commentaries that cover the third chapter of Genesis, or in Ambrose's *De paradiso*), and so it is perhaps the case that Bernard was drawing on the flora of the *De ave phoenice*, viewed through the lens of the later epic tradition, to endow his creation grove with a special numinous aura. It seems more likely that Bernard would have been familiar with this combination of plants from the poetry of the Christian *auctores*, which was widely available in verse anthologies, than through the most up-to-date Italian medical texts, with which he demonstrates no obvious familiarity elsewhere in the *Cosmographia*.[77]

To say this is not, however, to deny the more general medical significance of the plants in the creation grove. This aspect of the grove comes into focus if Bernard's account is set against a more explicitly theological tradition of

---

74 For the Horatian allusion, see *Cos.*, 22.

75 Palmer, "Plant Names", 44.

76 Lactantius, *De ave phoenice*, 83–87, in *Minor Latin Poets*, ed. and trans. J. Wright and A. M. Duff (Cambridge, MA: Harvard University Press, 1934), 656. Compare Victorius, *Alethia* II.236–39, ed. Martin, 137 (cinnamon, amomum, spikenard), and Dracontius, *De laudibus Dei* I.324–28, ed. Vollmer, 40 (cinnamon, spikenard, amomum). The biblical flora of the *hortus conclusus* (Song of Songs 4:14) are spikenard, cinnamon and saffron.

77 For the transmission of the biblical epics of late antiquity, see Irvine, *Making of Textual Culture*, 364–70; and M. Lapidge, "Versifying the Bible in the Middle Ages", in *The Text in the Community: Essays on Medieval Works, Manuscripts, Authors, and Readers*, ed. J. Mann and M. Nolan (Notre Dame, IN: University of Notre Dame Press, 2006), 11–40.

speculation about the nature of human physiology before and after the Fall. For Augustine, whose extensive discussions of the Fall made him the principal authority on this subject, man's consumption of the forbidden fruit from the Tree of Knowledge acted "to violate and corrupt the ordered integrity of his nature".[78] Although Augustine considered the Fall to have been a corruption of both body and soul, he was concerned almost exclusively with the spiritual ramifications, such as the 'depression' of free will. Such concerns continued to be of primary importance in the twelfth century.[79] However, this period also witnessed a new interest in the question of man's bodily debilitation. Augustine had thought that although the body of fallen man was of the same genus as Adam's, it was inferior because it would necessarily expire, whereas before it had been sustained by the Tree of Life, which produced "a secret inbreathing of health".[80] Where Augustine had posited a miraculous explanation for the resilience of man's pre-lapsarian condition, certain twelfth-century thinkers began to explore a more naturalistic account of the Fall.

They were assisted in this by contemporary medical theory, in particular the newly revitalised concept of the *complexio*, a term introduced by the medical translations produced in southern Italy in the late eleventh century.[81] In this theological context, it could be argued that man's original, well-balanced complexion had been critically destabilised by the Fall. Anselm of Laon (d. 1119) was an early proponent of this idea. He argued that "when Adam broke God's commandment his complexion was debilitated and corrupted by this transgression, as a punishment for his sin", and that it was through the inheritance of

---

78 Augustine, *De Genesi contra Manichaeos* II.9.12, ed. D. Weber (Vienna: Österreichische Akademie der Wissenschaften, 1998), 133.38–42: "Praecipitur ergo illi, ut *de omni ligno quod est in paradiso* [Gen. 2:16] edat, ex ligno autem in quo est dinoscentia boni et mali non edat, id est non sic eo fruatur, ut ipsam ordinatam integritatem naturae suae quasi manducando violet atque corrumpat."

79 Colish, *Peter Lombard*, I: 353–97.

80 Augustine, *De Genesi ad litteram* VII.5.11, in *Le Genèse au sens littéral en douze livres*, ed. and trans. P. Agaësse and A. Solignac, 2 vols. (Paris: Desclée de Brouwer, 1972), II: 26: "Illud quoque addo quamquam corporalem cibum, talem tamen illam arborem praestitisse, quo corpus hominis sanitate stabili firmaretur, non sicut ex alio cibo, sed nonnulla inspiratione salubritatis occulta"; and see also VI.26.37, ed. Agaësse and Solignac, I: 502: "Animale est enim et hoc corpus, sicut et primi hominis fuit; sed hoc iam in ipso animalis genere multo est deterius: habet enim necessitatem moriendi, quod illud non habuit." For the developments in Augustine's exegesis, see Y. K. Kim, *Augustine's Changing Interpretations of Genesis 1–3: From "De Genesi contra Manichaeos" to "De Genesi ad litteram'"*(Lewiston, NY: Edwin Mellen, 2006).

81 On the popularisation of the term *complexio*, through its use in Constantine the African's *Pantegni* and the *Isagoge* of Iohannicius, see Jacquart, "De *crasis* à *complexio*".

the corrupted particles of Adam's body that original sin had been transmitted.[82] This idea of punishment by complexional debilitation was developed in the 1150s by Hildegard of Bingen, whose theology was more deeply conditioned by medical thought. In her *Causae et curae*, Hildegard specified that man's pre-lapsarian complexion had been sanguine, but that it had been corrupted at the moment of disobedience by the coagulation in his blood of black bile, the humour associated with earth and therefore sin, and that this had precipitated the permanent melancholy of the human condition.[83] Here and elsewhere in her *Causae et curae*, Hildegard imagined an instantaneous act of corruption at the very moment ("in ipso momento") that Adam had eaten from the Tree of Knowledge. Occasionally, though, she suggested that it might have been the environment of paradise itself that sustained man's balanced complexion, and thus that his corruption could have been a result of his departure from that environment.[84]

This was certainly the opinion of William of Conches, whose *Dragmaticon philosophiae* (1144–1149) proposed not an abrupt discontinuity in human nature, but a gradual deterioration of physiology in response to the inferior environment outside Eden. Like Hildegard, William held that the ideal sanguine complexion had prevailed in pre-lapsarian man, but he described man's loss of natural heat and moisture in more starkly naturalistic terms. He wrote of Adam that:

> Postquam ab amoenitate paradisi expulsus in valle lachrimarum et miseriae in labore manuum suarum coepit vesci pane, suo labore vigiliis ieiuniis cepit desiccari atque naturalis calor extingui. Similiter ex intemperie aeris, ex qualitate cibi et potus. Omnes igitur ex eo nati, utpote ex corrupto, sunt corrupti.[85]

82 Anselm of Laon, *Sententie divine pagine* IV, in *Anselms von Laon systematische Sentenzen*, ed. F. Bliemetzrieder (Münster: Aschendorff, 1919), 33: "Quod Adam transgressus est mandatum Dei, ex illa transgressione complexio fuit debilitata et corrupta in pena peccati, et omnes partes eius debilitate sunt. Cum autem omnes essemus in lumbis Ade, particule ab eo in nos fluxerunt, ab eo etiam in nos fluxit corruptio, que appellatur peccatum. Id est, pena peccati, et originalis, quia ab origine fluit."

83 Hildegard of Bingen, *Cause et cure* II, ed. L. Moulinier (Berlin: Akademie, 2003), 183.8–11: "Cum enim Adam divinum preceptum prevaricatus est, in ipso momento melancolia in sanguine eius coagulata est, ut splendor recedit, cum lumen extinguitur, et ut stuppa ardens et fumigans fetendo remanet."

84 See, for example, *Cause et cure* II, ed. Moulinier, 59.5–6: "Si enim homo in paradyso mansisset, in imutabili et perfecto statu perstitisset."

85 William of Conches, *Dragmaticon philosophiae* VI.13.2–3, ed. Ronca, 227.11–16. Compare the more elliptical statement in William's earlier *Philosophia* IV.30, ed. Maurach, 103:

> After he had been driven from the pleasant land of paradise and began to eat bread by the labour of his hands in the valley of tears and misery, his body began to dry out from this labour as well as from the deprivations of food and sleep, and his natural heat began to fade away. Similarly, he was affected by the intemperate weather and the quality of his food and drink. Thus all who are born from him are corrupt, since they come from one corrupt.

Whereas Anselm and Hildegard used complexion theory to explain the instantaneous corruption of human nature posited by Augustine, William suggested that man's bodily debilitation was in fact entirely the result of the adverse environmental conditions encountered outside Eden. Instead of being sustained by a "secret inbreathing of health", miraculously transmitted by the Tree of Life, man's sanguine complexion (the ideal combination of moisture and heat) was sustained by the ideal environment of Eden, outside of which man "began to dry out... and his natural heat began to fade away".[86] The expulsion from Eden was of course an act of God, but for William the corruption of man's bodily nature was not itself a divine act, but a rationally explicable consequence of the adverse conditions outside Eden—of the hostile climate, the deprivations of food and drink, and the demands of manual labour.

William's approach here is entirely in keeping with a tendency among twelfth-century philosophers to minimise appeals to divine agency in explanations of change in nature.[87] The same tendency is evident in the *Cosmographia*, where Bernard Silvestris distinguishes the unfolding of creation "through the careful ordering of successive causes" ("dispositis causarum sibi succedentium ordinibus", I.4.6.8–9) from the initial "miraculum" of creation, in which the "elementorum structura" was established (I.2.12.1–2). Although Bernard does not refer directly to the effects of the Fall, it can be inferred from his account of human nature and the characteristics of Gramision that, like William, he considered man's expulsion from paradise to have been the cause of his bodily degradation. As we saw in the previous chapter, Bernard thought man's body to

---

"Verbi gratia, homo naturaliter calidus est et humidus et inter quattuor qualitates temperatus, sed quia corrumpitur natura, contingit illas in aliquo intendi et remitti."

86 Constantine the African, *De communibus locis* [=*Pantegni theorica*] I.20–23, in *Opera*, II: 17–19, had discussed the mutability of complexions in response to *regio*, *aetas*, *sexus* and *consuetudo* (lifestyle). Gregory, *Platonismo medievale: studi e ricerche* (Rome: Istituto storico italiano per il Medio Evo, 1958), 122–50, emphasises the naturalism of William's reading of Genesis; Elford, "William of Conches", 324–25, contrasts his account of Eden with Hildegard's.

87 See above, Chapter 2, 50–51.

consist of a well-balanced combination of humours, but to be susceptible nonetheless to the elemental destabilisation of Silva, who had not been entirely purged from the sublunary world.[88] The recollection of medical and astrological knowledge instilled by Physis and Urania is intended to offer some alleviation, but so too, I would argue, is the environment of Gramision—at least for the short time that man remains in the grove. For Gramision is perhaps the one place in the world that suffers none of the effects of elemental strife ("de elementorum intensione nichil in se suscipiens", II.9.3.5). And indeed so clear is the water flowing through the grove's stream that it is as if it had cast off its corporeity and returned to a state of ethereal purity ("etheree… puritatis", II.9.5.4).

These characteristics are reflected in the flora of Gramision, where there grows "whatever prevents disease and promotes health" ("quicquid occurit morbis, quicquid conciliat sanitatem", II.9.3.1). The plant catalogue, then, represents merely a sample of the full range of medicinal plants that grow within the paradise grove. Jolivet argued that there was a special gynaecological emphasis to Bernard's selection. However, many of the plants in question also had other medicinal uses.[89] And even if Bernard did have gynaecological applications in mind, his choice of plants might simply evoke certain treatments for the pains of childbirth, in the time before Eve and her descendants were to have such sorrows multiplied by the Lord (Gen. 3:16). Equally, the warm and moist climate of Gramision, which Jolivet associated with the optimum climatic conditions for conception, might instead refer to the exegetical commonplace that God created the world in springtime, the defining characteristics of which were heat and moisture.[90] Bernard's Gramision enjoys "perpetual spring" ("ver… perpetuum", II.9.3.7), and this may relate more specifically to William of Conches's contention (quoted above) that it was the favourable climate of Eden that helped man to maintain his ideal sanguine complexion,

88 See above, Chapter 2, 75–77.

89 Alternative medicinal uses for some of Bernard's plants are discussed, for example, in Pliny, *Historia naturalis* XII.26 (nard), 28 (amomum), 42 (cinnamon), 43 (cassia). Bernard's knowledge of the unusual *galanga* (= galingale, not Wetherbee's galbanum) may derive from Salernitan gynaecological texts, but more likely from Odo of Meung's late-eleventh-century botanical treatise, *De viribus herbarum* 50, ed. L. Choulant (Leipzig: Voss, 1832), 117, where *galanga* is described as an aid to digestion. For Bernard's botanical sources see above, Chapter 2, 78.

90 J. Pépin, "Exégèse de *In principio* et théorie des principes dans l'*Exameron* (I.4.12–16)", in *Ambrosius Episcopus*, ed. G. Lazatti, 2 vols. (Milan: Vita e Penserio, 1976), I: 427–82 (441–54) (repr. in Pépin, *De la philosophie ancienne à la théologie patristique* (London: Variorum Reprints, 1986), XV).

which was also characterised by a combination of hot and moist elemental qualities.

The climate of Gramision is then akin to that of the superlunar world. The grove is a haven from the turmoil of the sublunar sphere, a place in which the reasoning powers of the human soul may function with a minimum of external interference. Such a link between climate and the *vita contemplativa* may also be found in Adelard of Bath's *Quaestiones naturales* (1111–1116), where one of the arguments advanced for the animation of the heavens is that the more purified climates of Earth (those closest in consistency to the heavens) are those that are most productive of philosophers—those humans whose reasoning corresponds most closely to the perfect rotations of the *anima mundi*.[91] In Bernard's superlunar world, Endelechia could ensure that a code of *hospitium* (I.2.15.15) was maintained, and that the malign influence of Silva would not disturb the harmony of the cosmos. Wetherbee has rightly noted that this idea of cosmic *hospitium* anticipates Bernard's lines on the Fall in the following verse chapter:[92]

> Hos, reor, incoluit riguos pictosque recessus
> *Hospes*—sed brevior *hospite*—primus homo.
> Hoc studio curante nemus Natura creavit—
> Surgit fortuitis cetera silva locis. (I.3.335–38)

> In this brightly tinted fertile haunt, I think, dwelt man
> At first, too briefly to be called a guest.
> With zealous care did Nature craft this grove, while woods
> In other quarters grew at random.

Whilst abiding by the code of *hospitium*, man enjoyed the privileges of inhabiting a terrestrial realm whose elemental conditions were akin to those of the tranquil superlunar world. However, because Adam and Eve violated the laws

91 *Quaestiones naturales* 74, in Adelard of Bath, *Conversatsions with his Nephew*, ed. Burnett *et al.*, 218–19: "Amplius: si in hac ipsa tenebrosa regione nostra ille locus philosophorum fertilior est, qui serenitate quadam a densa fece elimatus purgatior est, qua fronte illam plagam celestem que quanto a nobis remotissima, tanto omni tali sorde intactissima est, animali motu expertem esse dicere sine summa insania quisquam potest?" The editors (234, n. 88) note a similar association between climate and intelligence in Cicero, *De natura deorum* II.xv–xvi (42).

92 *Cosmographia*, trans. Wetherbee, 148, n. 53.

of hospitality, by eating fruit that was forbidden them, they were exiled from this carefully prepared *nemus* into the random *silva*.

One might read this distinction between the two types of woodland as simple rhetorical variation, announcing the catalogue of extra-paradisiacal woods enumerated in the following verses, which range from the Pierian *nemus* to the *silva* of the Ardennes (ll. 339–54). However, given the philosophical charge of the term in the *Cosmographia*, it seems reasonable to assume that *silva* here also denotes the turbulent material world, "the image of the ancient chaos" ("tumultus instar veteris", II.5.21.4).[93] This couplet may be read, then, as a natural scientific gloss on the *terra maledicta* (Gen. 3:17) in which fallen man is condemned to live. Whilst in Gramision, man's soul is less disturbed by Silva, who impairs reasoning in the fallen world.[94] It is only with his death that man may attain once again the status of the *hospes* that he forfeited at the Fall. Only then, Noys declares, will man leave the tottering structure of his body and return to the stars—"iam non incognitus hospes" (II.10.53).

Bernard exploited the dual meaning of *silva* as both woodland/wilderness and primal matter to suggest the special status of Gramision as a haven from elemental discord. It is hardly surprising, then, that of the various topoi of inaccessibility available to speculative geographers of Eden, Bernard chose not the unscalable mountain (Ezek. 28:13–14), or the walls of fire (Gen. 3:24), but "the surrounding woodland, which prevented open access" ("silva lateraliter circumplectens … [quae] communes arcebat ingressus", II.9.5.8–9). This topos has no scriptural authority, but it does appear in the description of Eden in the late antique biblical epic *Alethia*, where the poet writes of "the lap [*gremium*] of the sacred grove [*nemus*], which woodland [*silva*] encircles."[95] Although comprising the same topographic features, this description has none of the philosophical charge of Bernard's lines. In the *Cosmographia*, the conventions of the *locus amoenus* are deployed to underline the elemental distinction between Gramision and the rest of the sublunar world. The theological implication is that the corruption of man's fallen body is a gradual process to be

---

93 Note also the connection between the random places ("fortuitis … locis", I.3.338) in which the *silva* grows outside Gramision and the random movements ("fortuitis … modis", I.1.26) of the elements within chaotic Silva at the opening of the work.

94 In the commentary on the *Aeneid* sometimes attributed to Bernard Silvestris, the commentator interprets the passage (VI.179–82) in which the Trojans enter *in antiquam silvam* to fell trees for the funeral pyre of Misenus as the uprooting of vice in preparation for the cultivation of virtue. See *Commentum*, ed. Jones and Jones, 61–63. For the debate over attribution see above, Chapter 1, 27–28.

95 Victorius, *Alethia* II.268, ed. Martin, 138: "At gremium sacri nemoris, quod silva coronat."

attributed to a change in environment, as in William of Conches.[96] Thus Bernard followed William in departing from the traditional, Augustinian idea, developed by Anselm of Laon and Hildegard of Bingen, that God punished man with the instantaneous debilitation of his bodily nature.

This, then, is the purpose of Bernard's fable of Gramision. It should not, I have argued, be understood in terms of the pagan fertility themes identified by Curtius, who wanted to see in the *Cosmographia* the revival of an ancient fertility cult, for which, as he himself admitted, "we have no other indications... in the Middle Ages, except in certain Grail romances".[97] The supposedly pagan features of Gramision adduced by Jolivet in support of this thesis—the grove as a fertile *locus amoenus* and a source of medicinal plants—point rather to the construction of Gramision as an *integumentum* of the Garden of Eden. What the fable of Gramision 'clothes' is a naturalistic understanding of the Fall of a kind advanced by William of Conches, and in keeping with a contemporary philosophical trend to limit appeals to divine agency in explanations of change in nature. The expulsion from Eden was of course an act of God, a suspension of the divine *fides hospitii*. However, the corrupting of man's bodily nature was not itself a divine intervention, but rather a consequence of the adverse environmental conditions outside Eden. William of Conches expressed this radical naturalism unambiguously, and was censured for it by William of Saint-Thierry, who thought the position entailed an unacceptable, Manichean division between divine and natural causation, according to which God's activity in the world would be reduced to nothing more than "the clashing of elements" (*concursus elementorum*) and "the tempering of nature" (*temperatura naturae*).[98] Bernard Silvestris instead deployed a myth simply to present,

96 William understood the trajectory of man's gradual bodily decline in terms of the cycle of cosmic creation and destruction determined by the passing of the Platonic world-year (the cycle of 15,000 solar years described by Macrobius, *Commentarii* II.10–11). Alluding to the Apocalypse and quoting Juvenal, William wrote (*Dragmaticon philosophiae* III.4.6, ed. Ronca, 67.54–56): "Unde credimus finem mundi instare, quia quantitates corporum imminutas, vitas abbreviatas videmus: terra malos homines nunc educat atque pusillos [*Sat.* XV.70]." Bernard had a similar conception of the cyclical nature of time (see *Cos.* I.4, with the commentary in Stock, *Myth and Science*, 157–62), and he saw a period of degeneration from the golden age to an age of iron: "Secularis illa continentia, purioribus ex auro iniciata principiis, paulatim degenerante materia, in ferrum visa est terminare" (II.11.9.8–9).

97 Curtius, *Europäische Literatur*, 120: "Wir kennen das im Mittelalter sonst nur aus den Andeutungen einiger Gralromane"; trans. Trask, *European Literature*, 112.

98 William of Saint-Thierry, *De erroribus Guilelmi de Conchis* 8, ed. Leclercq, 368.329–35: "Ubi in altero quidem stultorum quorumdam philosophorum videtur sententiam sequi, dicen-

rather than insist upon, this philosophical revision of the patristic theological legacy. As in his passages on the Trinity and creation, the philosophical is placed in juxtaposition to the theological, and it is left to the reader to resolve any tensions that may result.

The Incarnation

The same integumental approach may be seen at work in our final example, which concerns the theological implications of astrology, specifically in relation to the Incarnation. The passage in question comes near the beginning of the catalogue poem, following some lines on the celestial hierarchy. Bernard is describing the nature and significance of the fixed stars:

Terrenis excepta super, substantia celi,
 Ut melior, cultu sic meliore fuit;
Scribit enim celum stellis, totumque figurat
 Quod de fatali lege venire potest.
Presignat qualique modo qualique tenore
 Omnia sidereus secula motus agat.
Preiacet in stellis series quam longior etas
 Explicet et spatiis temporis ordo suis:
Sceptra Phoronei, fratrum discordia Thebe,
 Flamme Phetontis, Deucalionis aque;
In stellis Codri paupertas, copia Cresi,
 Incestus Paridis Ypolitique pudor;
In stellis Priami species, audatia Turni,
 Sensus Ulixeus, Herculeusque vigor;
In stellis pugil est Pollux, et navita Tiphis,
 Et Cicero rhetor, et geometra Thales;
In stellis lepidum dictat Maro, Milo figurat,
 Fulgurat in Lacia nobilitate Nero;
Astra notat Persis, Egyptus parturit artes,
 Grecia docta legit, prelia Roma gerit;
In causas rerum sentit Plato, pugnat Achilles,

tium nihil prorsus esse praeter corpora et corporea, non aliud esse Deum in mundo quam concursum elementorum et temperaturam naturae, et hoc ipsum esse animam in corpore, in altero manifestus Manichaeus est, dicens animam hominis a bono Deo creatam, corpus vero a principe tenebrarum."

Et prelarga Tyti dextera spargit opes;
Exemplar speciemque dei virguncula Christum
Parturit, et verum secula numen habent;
Munificiens deitas Eugenium comodat orbi,
Donat et in solo munere cuncta semel.
Sic opifex, ut in ante queant ventura videri
Secula, sidereis significata modis. (I.3.31–58)

Drawn high 'bove earthly things, the finer substance
Of the heavens is clothed in finer garb;
For here Noys scribes the sky with stars, and figures
All fate's law may bring to pass.
She shows the means and path by which the stars
In motion drive all worldly things.
The stars portend a course that distant ages
And time's measured train unfold:
Phoroneus' rule, the brothers' Theban strife,
Blazing Phaeton, Deucalion's flood.
Instarred are Codrus' want and Croesus' wealth,
Lustful Hippolytus, Paris chaste.
Instarred are Priam's splendour, Turnus' boldness,
Odyssean cunning, Hercules' strength.
Instarred are Pollux the boxer, Tiphys the helmsman,
The orator Cicero, geometry's Thales.
Figures and words grace Myron and Vergil,
While Nero gleams in Latian pomp.
Persia has star-maps, Egypt the arts,
Greece sagely reads, Rome wages war.
Plato fathoms creation, Achilles contests,
Titus' bountiful hand spreads wealth.
A virgin begets Christ, likeness and figure
Of God, and Earth receives the divine.
God graciously gives to the world Eugenius,
And grants in this one gift all at once.
Thus wrought the creator, that ages to come may be seen
In advance, signalled by starry ciphers.

The theological implications of this passage have been construed in various ways. The earliest modern critics of the *Cosmographia*, the Benedictine authors of the *Histoire littéraire de la France*, were appalled by "the insane doctrine contained in these verses"—in general the deterministic outlook of the passage and in particular the couplet implying the astrological predictability of the Incarnation.[99] Gilson, in his polemic against this kind of reading, skated over the problematic context of the couplet, commenting simply that the poet "teaches moreover that Christ is the true God, born of a virgin".[100] But for Curtius, reacting in turn to this defence of Bernard's Christianity, the passage was further evidence of the "pagan humanism" of an author who seems "to have attached more importance to Phoroneus, Paris, Pollux and their colleagues than to the Christian revelation and its prefiguration in the Old Testament".[101]

On first inspection, Curtius's comment about the balance of Bernard's interests is borne out by the text of the poem, which devotes fourteen lines to the peoples of pagan antiquity and only four to the Christian era. However, a closer look at this and the other lists in the catalogue poem reveals a more complex system behind Bernard's arrangement of his material. Christine Ratkowitsch has described the catalogue poem as a kind of versified *mappa mundi*, structured by the same spatio-temporal principles that informed the pictorial and textual representation of the world in contemporary maps.[102] Thus, in the various geographical and historical surveys of which the catalogue poem is composed, Bernard tended to move from East to West and through the three *tempora* of providential history—*ante legem, sub lege* and *sub gratia*.[103] Two

99 *Histoire littéraire de la France*, XII: 271: "... la doctrine insensée contenue dans ces vers ... [ce] dernier distique surtout fait horreur."

100 "La cosmogonie", 20: "[Bernard] enseigne en outre que le Christ est le vrai Dieu, né d'une vièrge."

101 *Europäische Literatur*, 117: "Der Noys und ihrem Dichter scheinen Phoroneus, Paris, Pollux und ihre Kollegen wichtiger gewesen zu sein als die christliche Heilsgeschichte und ihre Präfiguration im Alten Testament"; trans. Trask, *European Literature*, 109–10.

102 Ratkowitsch, *Cosmographia*, 50–51; and for these principles in general, see A. Scafi, "Defining *Mappaemundi*", in *The Hereford World Map: Medieval World Maps and their Context*, ed. P. D. A. Harvey (London: British Library, 2006), 345–54.

103 This structure can also be seen in Baudri of Bourgueil's versified ecphrasis of the tapestry depicting world history that decorated the bedchamber of his patron, Countess Adela of Blois; composed *c.* 1100, his poem may have been known to Bernard. See C. Ratkowitsch, *Descriptio Picturae: die literarische Funktion der Beschreibung von Kunstwerken in der lateinischen Großdichtung des 12. Jahrhunderts* (Vienna: Österreichische Akademie der Wissenschaften, 1991), 34, n. 39. Of course behind both the pictorial and the poetic sources adduced by Ratkowitsch lies the tradition of universal chronicles, from Orosius in late antiquity through to Marianus Scotus, Sigebert of Gembloux and Hugh of Flavigny in the eleventh century. Further research is required to determine the precise sources of the historical data in Bernard's poem.

good examples of this spatio-temporal dynamic in Bernard's poem are his mountain and river catalogues. The first, which lists twenty-four mountains (I.3.175–98), begins with Olympus, the home of the gods, then moves through Lebanon with its cedars (*cf.* Song of Sol. 5:15), to Sinai, where the law was given to Moses, and on to Monte Gargano, associated in the Middle Ages with the cult of St Michael, before finishing with the Western European Alps. The same dynamic is evident in the river catalogue (ll. 237–64), which begins with the Tigris and Euphrates (the rivers that flow out of Eden (Gen. 2:14), but here coupled to classical allusions to the *tempus ante legem*), then moves through the biblical rivers of the Abana (2 Kings 5:12), the Shiloah (John 9:7–11) and the Jordan, before moving once again to Western Europe, finishing in Gaul with the Seine and the Loire, the river of the poet's own city.[104]

As Ratkowitsch's rich commentary makes clear, many of the passages in the catalogue poem are not simply collections of inert information, but sets of allusions and prompts demanding of the reader a knowledge of Scripture and the *auctores* in order to tease out a deeper meaning. Bernard's allusions, like those in John of Salisbury's *Policraticus*, are "mnemonic devices aiding inquiry and points to which ideas can be related by similarity, dissimilarity or contrast".[105] This is certainly the case in the astrological catalogue. Here, too, there is a pseudo-chronological development from Phoroneus, the legendary first lawmaker of the Greeks (see Isidore, *Etymologiae* V.1.1), through to Christ and Eugenius III, the pope of Bernard's day. But more important are the qualities associated with these exemplary figures. They are a mixture of positive and negative *exempla*, or, as Wetherbee calls them, "prototypes of wisdom, beauty, and strength, and of vice, vanity, and treachery".[106] Thus, there are examples from pagan antiquity of civil harmony and discord, in the contrasting pair of Phoroneus and the warring brothers of Thebes; of incontinence and chastity, in the coupling of Paris and Hippolytus; and of the beauty of artistic creation, represented by Vergil in poetry and Myron in sculpture, contrasted with the decadent grandeur of Nero.[107] However, all the individual pagan *artes* and *virtutes*—Plato's philosophical knowledge of God, the heroism of Achilles,

---

104 For commentary, see Ratkowitsch, *Cosmographia*, 62–65, 67–71; and Wetherbee, *Platonism and Poetry*, 166–72; and on the Loire, see above, Chapter 1, 19–20.

105 P. von Moos, "The Use of *Exempla* in the *Policraticus* of John of Salisbury", in *World of John of Salisbury*, ed. Wilks, 207–61 (232).

106 Wetherbee, *Platonism and Poetry*, 167.

107 See Suetonius, *Nero* 20–31, 42, and Conrad of Hirsau's *Dialogus super auctores* (after 1124/25), which praises Lucan's ironic portrait of the emperor (*Accessus ad auctores*, ed. Huygens, 110).

and the generosity of Titus (Suetonius, *Titus* 7)—are consummated in the figure of Christ, for only with the Incarnation does "earth receive the divine" ("secula numen habent", l. 54). Christ, then, is the glorious consummation of Bernard's catalogue, not, as Curtius thought, a barely significant coda.

Granted the importance Bernard attached to Christ in the unfolding of the cosmos, there remains the question of the orthodoxy of his portrayal of the Incarnation as an astrologically foreseeable event. Astrology had been widely condemned in patristic theology as a pernicious feature of pagan culture. For astrological prediction presupposed a fixed and unalterable movement of the stars that determined the course of human affairs, and this implied a succession of necessary causes that Christian theologians saw as an affront both to human free will and divine omnipotence.[108] A particular challenge to this view was presented by the Gospel text of Matthew 2:2 on the "star in the East" that alerted the Magi to the birth of Christ. In patristic homilies on Matthew and sermons for the Epiphany it was made clear that the Magi *were* astrologers, but respectable ones who had set aside their superstitious practices. Following the text of the Gospel itself, theologians asserted that the star of Bethlehem had not been forever present in the heavens, but that it had "appeared" (*apparuit*) and then led the Magi to Bethlehem, where it came to rest (Matt. 2:7–9). It was, therefore, a miraculous creation brought into being exclusively to indicate the birthplace of Christ, not part of an ever-present celestial configuration that might be read as a horoscope of the Incarnation.[109]

A sense of the continuity of patristic hostility towards astrology may be discerned from the numerous condemnations of the practice in early medieval canon law, and in Isidore of Seville's *Etymologiae*, which distinguished between natural and superstitious astrology, the first consisting of the legitimate study of the course of the sun, moon and stars (useful for weather forecasting, farming, navigation and *computus*), and the second involving the reprehensible activity of casting horoscopes to determine human characteristics.[110] There appears to have been a certain softening in this attitude towards astrology from

108 See the critique of astrology in Augustine, *De civitate Dei* V.1–10.

109 For the patristic sources, see V. I. J. Flint, *The Rise of Magic in Early Medieval Europe* (Oxford: Clarendon Press, 1991), 368–69.

110 Isidore of Seville, *Etymologiae* III.27: "Astrologia vero partim naturalis, partim superstitiosa est. Naturalis, dum exequitur solis et lunae cursus, vel stellarum certas temporum stationes. Superstitiosa vero est illa quam mathematici sequuntur, qui in stellis augurantur, quique etiam duodecim caeli signa per singula animae vel corporis membra disponunt, siderumque cursu nativitates hominum et mores praedicare conantur." For the social context of astrology in the early Middle Ages, see Flint, *Rise of Magic*, 92–101, 128–46.

the ninth century, and this coincided with a rehabilitation of the Magi as legitimate astrologers.[111] However, even in these more favourable assessments there was no suggestion that the Magi had relied on a horoscope to predict the birth of Christ, but rather it was thought they had used Old Testament learning, for example the prophecy of Balaam (Num. 24:17) that a new star would announce the birth of the Messiah.[112] The first suggestion of an astrological explanation of the Star of Bethlehem came with the Latin translations, in 1133 and 1140, of Abu Ma'shar's *Introductorium maius*, which transmitted from the Arabic an ancient horoscope predicting that in the first decan of the constellation Virgo a virgin would arise feeding a boy from her breast—and this boy, according to Abu Ma'shar, "is known to certain nations as Jesus."[113] In his *De essentiis* Hermann of Carinthia quoted this passage from his translation of Abu Ma'shar, explicitly linking it with the Magi, and using it as a vindication of the astrologer's art:

> Quem, opinor, ex hac lectione instructi, visa stella eius, et Magi cognoverunt. Unde presertim Hebreorum cecitas hic arguitur, cum, etiam in nature speculatione seculorumque serie, vel barbare nationi veritas Ihesu Christi prenotata fuerit. Planius quippe rem omnem vidit astrologus. Quam mirum quod preter naturam futurum videbat, effari posset!
>
> In my opinion the Wise Men also recognized the boy, being informed by reading this, and having seen his star. Thus especially the blindness of the Hebrews is shown here, since even in natural speculation and the order

111 Flint, *Rise of Magic*, 364–75.

112 See pseudo-Bede, *In Mattaei evangelium expositio* I.2, *PL* 92, 9–132 (12D–13A): "*Ecce Magi ab Oriente*, usque, *Venimus adorare eum*. Magi non propter magicam artem sic nominantur, sed propter aliquam philosophiam, in qua successores Balaam esse creduntur, et propter prophetiam eius incitati, visa nova stella natum regem quaerebant, qui confingere omnes duces aliengenarum prophetatus est ex divino eorum."

113 *Introductorium maius* VI.1, ed. Lemay, VIII (trans. Hermann of Carinthia): 101.256–64 (with Hermann's interpolations in parentheses): "Virgo signum fertile, bipertitum, triforme. Oritur in primo eius decano ut Perse, Caldei et Egyptii, omniumque duces Hermes et Astalius a primeva etate docent, Puella cui persicum nomen *Seclios Darzama*, arabice interpretatum *Adre nedefa* i.e. virgo munda (puella dico Virgo Immaculata), corpore decora, vultu venusta, habitu modesta, crine prolixo, manu geminas aristas tenens, supra solium auleatum residens, puerum nutriens ac iusse pascens in loco cui nomen Hebrea, puerum inquam a quibusdam nationibus nominatum IHESU significantibus *Eiza* (quem nos grece CHRISTUM dicimus)." For John of Seville's translation, see *Introductorium maius* VI.1, ed. Lemay, V: 224.429–36.

> of the ages, the truth of Jesus Christ was in fact first known by a foreign nation. For the astrologer saw the whole situation more clearly. How wonderful it is that he could announce a wonder which he saw would happen in the future against the laws of nature![114]

Hermann's plea for the place of astrology in Christian learning was part of a wider attempt in the mid-twelfth century to rehabilitate astrology as a respectable scientific discipline—a project in which the *Cosmographia* itself was, as we have seen, heavily invested.[115] Another important work in this regard was Raymond of Marseille's *Liber cursuum planetarum* (1141), which argued for the compatibility of Christianity and astrology, and provided, as Hermann had, a biblical warrant for the astrologer's art, according to the authority of Gen. 1:14, that the newly created heavenly bodies "may be used for signs and seasons" ("et sint in signa et tempora").[116]

However, this attempt at a theological rehabilitation of astrology was controversial and many of the old arguments against the discipline resurfaced in this period. Thus Abelard's *Hexameron*, probably written in the early to mid-1130s, before Raymond's work but nonetheless responding to the contemporary vogue for astrology, contains a different interpretation of the *signa* of Gen. 1:14. There, following the Isidorian distinction between natural and superstitious astrology, Abelard wrote that the heavenly bodies were signs of events occurring according to the natural order of things—in the way that death follows poisoning or rain follows thunder—but that they could not be used to foresee contingent events.[117] Since astrology was a branch of physics, which is

---

114 *De essentiis* 59rGH, ed. and trans. Burnett, 82–83.

115 M.-T. d'Alverny, "Astrologues et théologiens au xiie siècle", in *Mélanges offerts à M.-D. Chenu, Maître en théologie* (Paris: Vrin, 1967), 31–50. New natural philosophical emphases are also evident in discussions of the Magi by Alan of Lille and Jacques de Vitry: see Ferruolo, *Origins*, 255.

116 *Liber cursuum planetarum* I.9, in Raymond of Marseille, *Opera Omnia I*, ed. M.-T. d'Alverny, C. Burnett and E. Poulle (Paris: CNRS, 2009), 140: "Quoniam stellas Omnipotentis provida dispositio mirabiliter in celo collocavit, ut non tantummodo sint nobis in tempora, verum etiam in signa, ut diximus, cum philosophi planetas in signis quasi animas in corporibus operari efficaciter atque mirabiliter pro sui Artificis mirifici mandato sentiant, ipsorumque planetarum significationes secundum signorum in quibus fuerint proprietates et complexiones variari affirment."

117 *Expositio in Hexameron* 193, in *Opera Theologica, V*, ed. Romig *et al.*, 47.1195–1203: "Naturalia vero futura dicuntur, que causam aliquam naturalem sui eventus habent, ut ex his que precedunt, tamquam quibusdam naturalibus sui causis contingere habeant, et sic eis adiuncta sint, ut vix aut numquam illis precedentibus hec inde contingere possit

the study of nature, how could it be used to predict things that lay beyond nature?[118] Similar arguments were advanced, at greater length, in John of Salisbury's *Policraticus* (*c.* 1159), which identified astrology as one of the courtly trifles of the time, and as a misguided product of the contemporary interest in *natura* and *ratio* that had led its exponents to deny human free will and God's control over the world.[119]

The doctrine of the astrological foreseeability of the Incarnation, that is the idea of a horoscope that might have predicted the virgin birth at a certain time and place, is an extreme example of the kind of determinism that Abelard and John condemned. For the miracle of the virgin birth would surely be the most clear-cut instance of contingency imaginable, and therefore the event furthest removed from the realm of the *naturalia* with which legitimate astrology was concerned. In Calcidian terms, the Incarnation was an *opus Dei* rather than an *opus naturae*, and indeed William of Conches used it as a paradigmatic example of Calcidius's first category.[120] Hermann of Carinthia agreed that the virgin birth was "preter naturam", but at the same time insisted that it had been foreseen "in nature speculatione". This was a radical and, in a sense, a paradoxical claim, for it implied that the scientist could exploit the study of nature to look behind the natural order, to perceive something of the *voluntas Dei* that had instituted, and would occasionally suspend, the laws of nature.

Discussion of this astrological forecast of the Incarnation appears in the brief preface to the *Experimentarius*, which of all the parts of that divinatory compilation is the one most likely to have been written by Bernard. It is grouped there with other terrestrial and celestial phenomena that God revealed through the stars.[121] Even if this is not Bernard's work, it would none-

---

impediri: velut mortis dissolutio futura in proximo post venenum exhaustum, vel pluvia post tonitrum, vel sterilitas terre post nimiam astuationem vel immoderatos imbres."

118 *Expositio in Hexameron* 198, ed. Romig *et al.*, 48.1227–31: "Cum enim astronomia species sit phisice, hoc est naturalis philosophie, quomodo per eam investigare valent quod ipsi quoque nature philosophi asserunt incognitum esse, hoc est ex nulla rerum natura prenosci posse?"

119 *Policraticus* II.18–26.

120 *Glosae super Platonem* 37, ed. Jeauneau, 69.4–7: "Et est opus Creatoris ... ea quae videmus fieri contra consuetum cursum naturae ut partus virginis, etc."

121 *Experimentarius*, praef., ed. Burnett, "What is the *Experimentarius* of Bernardus Silvestris?", 123: "De qualitatibus etiam hominum necnon et aliorum animalium, sanitatibus et egritudinibus, statu et prosperitate, telluris fertilitate, maris tempestate, aeris qualitate, institorum negociacione et peregrinantium reditu diligenter intendentibus [Deus] revelat evidenciam, ut patuit tribus magis nativitas magni prophete veritatis Ihesu." Burnett concludes that there is no good evidence for attributing to Bernard any part of

theless seem reasonable to assume that when he wrote his lines on Christ in the *Cosmographia* he had in mind the forecast recently transmitted through the translations of Abu Ma'shar.[122] In the same poem in which Christ appears, Bernard writes explicitly about the influence of the stars on human affairs, and also about man's capacity to see the future in the heavens. By means of the star-studded firmament, he writes, God "prefigures all that may come to pass through fated law".[123] And in the long extract quoted above, Bernard explains that the heavens were created "so that ages to come might be beheld in advance" ("ut in ante queant ventura videri | Secula", ll. 57–58). As Dronke has rightly observed, this does not means that all the events of history are prefigured in the stars, but that certain characteristics are innate both to individuals and to collective peoples, of which Bernard lists some famous examples in his poem.[124] The characteristic of Christ is his divinity, and Bernard did not skirt the difficult issue that according to his metaphysics this meant that the Incarnation must have been foreseeable through astrology. However, as with his treatment of creation and the Fall, Bernard left this scientific thesis hanging in an uncertain relationship with theological tradition—here with the Christian conception of providential history that informs the catalogue poem in which this controversial thesis is imbedded.

This is the sense in which Bernard's astrological catalogue is a fable or covering (*integumentum*) behind which philosophical and theological truths may be discovered. For Bernard supported the notion that the Incarnation was astrologically foreseeable, but he attenuated an objection that this was an

---

the *Experimentarius* beyond this preface. M. T. Donati, "Metafisica, fisica e astrologia nel xii secolo: Bernardo Silvestre e l'introduzione *qui celum* dell'*Experimentarius*", *Studi Medievali* 31 (1990): 649–703, ascribes to him an additional prefatory passage from the collection. But the argument does not convince, for this other passage advances a miraculous account of the Star of Bethlehem that contradicts the naturalism both of the first passage and of the *Cosmographia*. See *Experimentarius*, ed. Burnett, 116 (italics added): "Quo placuit autem creatori tempore prodigia et signa faciens mundo innotuit, primum Iudeis, et demum toti mundo, sicut omnipotens quantum vouluit se manifestavit, *et contra et super posse planetarum suum posse demonstrans*, in suo etiam ortu *novam stellam* gentibus revelavit..."

122 This was one of the most influential passages of the *Introductorium maius*. The horoscope is cited, for example, in a twelfth-century *Microcosmographia*, perhaps written at Reims, which appears to make no other use of Abu Ma'shar: see J. R. Williams, "The *Microcosmographia* of Trier MS 1041", *Isis* 22 (1934): 106–35 (110).

123 *Cos.* I.3.33–34: "Scribit enim celum stellis, totumque figurat | Quod de fatali lege venire potest."

124 *Fabula*, 141–43.

affront to divine omnipotence by placing this notion within a narrative that revealed the role of the divine hand in the shaping of world history, culminating in Christianity. Curtius remarked on the absence from the astrological catalogue of any reference to the "prefiguration" of Christ in the Old Testament.[125] Bernard instead emphasised the role of the firmament that "prefigures" (l. 33) the future. He worked with an astrological, as opposed to a typological, notion of *figura*, and throughout the catalogue poem used this notion to evoke a sense of the divine plan woven into the texture of human history.

The Function of *integumentum*

I have proposed in this chapter that Bernard Silvestris adopted an integumental approach to theological questions, whereby Christian doctrine was 'clothed' in the garments of philosophy and allegory. Bernard nowhere makes clear the purpose of his integumental approach, omitting any guidelines for interpretation of the kind that Alan of Lille would later include in the preface to his *Anticlaudianus*. However, some light may be shed on his motives by considering the definition of *integumentum* presented in the commentary on Martianus Capella sometimes attributed to Bernard.[126] In the introduction (*accessus*) the commentator defines *integumentum* as a specific form of figurative discourse:

> Figura autem est oratio quam involucrum dicere solent. Hec bipertita est: partimur namque eam in allegoriam et integumentum. Est autem allegoria oratio sub historica narratione verum et ab exteriori diversum involvens intellectum, ut de lucta Iacob. Integumentum vero est oratio sub fabulosa narratione verum claudens intellectum, ut de Orpheo. Nam et ibi historia et hic fabula misterium habent occultum, quod alias discutiendum erit. Allegoria quidem divine pagine, integumentum vero philosophice competit.

> Figurative discourse is a mode of discourse that is called 'a veil'. Figurative discourse is twofold, for we divide it into allegory and *integumentum*. Allegory is a mode of discourse that covers under a historical narrative a true meaning that is different from its surface meaning, as in the case of Jacob wrestling with the angel. An *integumentum*, however, is a mode of

125 *European Literature*, 109–110.

126 For the debate over attribution see above, Chapter 1, 27–28.

> discourse that covers a true meaning under a fictitious narrative, as in the case of Orpheus. For history in the former case, and fiction in the latter, both contain a profound hidden truth, which will be explained elsewhere. Allegory pertains to Holy Scripture, but *integumentum* to philosophical scripture.[127]

Here we have a comparison and contrast between *integumentum* and *allegoria*. Both contain truths hidden behind the surface narrative. They differ in that *integumentum* is a fabulous narrative that is not literally true, whilst allegory is a biblical narrative that is literally true. The blurring of this distinction, by treating Scripture as *integumentum*, was one of the controversial aspects of William of Conches's philosophy, for William of Saint-Thierry accused him of heretically denying the literal truth of Eve's creation from Adam's rib.[128] And elsewhere in the aforementioned Martianus commentary, the author attacked William of Conches's thesis that the waters above the firmament described by Gen. 1:6–7 were a physical impossibility—for they required the suspension of a heavier element (water) over a lighter one (air)—and that they should not therefore be understood literally. Citing a number of biblical and patristic authorities, the Martianus commentator reasserted the traditional argument for a literal reading of these verses from the *divina pagina*.[129]

William of Conches treated Scripture as *integumentum* in order to bring it into agreement with the truths of natural philosophy. But Bernard was more equivocal in his handling of figurative discourse and truth content: he left it to his readers to determine the relationship between his philosophical fable and the doctrines of revealed religion. Thus, the critical debate about the theology of the *Cosmographia*—whether it is a Christian allegory (Gilson) or a pagan fable (Curtius)—has enacted a dialectic of interpretation that was an intended

127 *Commentum in Martianum* 2.70–78, in *The Commentary on Martianus Capella's "De nuptiis Philologiae et Mercurii" Attributed to Bernardus Silvestris*, ed. H. J. Westra (Toronto: PIMS, 1986), 24, 45.

128 *De erroribus* 9, ed. Leclercq, 368.336–39: "In creatione vero mulieris palam omnibus legentibus est, quam stulte, quam superbe irridet historiam divinae auctoritatis, scilicet excostasse Deum primum hominem ad faciendam de costa eius mulierem." Compare William of Conches, *Philosophia* I.13, ed. Maurach, 38: "Non enim ad litteram credendum est deum excostasse primum hominem." William thought that this verse symbolised the inferior complexion of woman, having been created *next to* the ideal location in which man was made.

129 *Commentum in Martianum* 8.410–71, ed. Westra, 186–88; and see further H. R. Lemay, "Science and Theology at Chartres: The Case of the Supracelestial Waters", *British Journal for the History of Science* 10 (1977): 226–36.

consequence of the work. Critics who have been sympathetic to the integumental character of the work have often assumed that Bernard used poetry to explore the kind of controversial doctrines for which his prose-writing contemporaries were condemned—for "what philosophy could not do, poetry might".[130] Yet such an interpretation ultimately relies on the same view of Bernard's myth making as that advanced by Gilson, who saw the "allegories borrowed from ancient poets" as a literary veneer beneath which it was possible (if only for the most learned) to discover a coherent message.[131] Many twelfth-century readers, by contrast, would have emphasised the multiplicity of possible meanings concealed beneath a literary *integumentum*. Thus in the Martianus Capella commentary, the *accessus* warns the student of the "double and even multiple meanings" ("equivocationes et multivocationes") to be found in the *De nuptiis*.[132] Likewise, in his commentary on Boethius's poetic version of the Orpheus and Eurydice myth, William of Conches justified his departure from the interpretation of Fulgentius with the well-known remark that "one should not worry, but rather rejoice about a diversity of interpretations."[133]

In imitation of Martianus and Boethius, Bernard adopted the mixed form of alternating chapters of verse and prose. And his work seems also designed to convey the same sense of depth and multiplicity of meaning that contemporary commentators found in the ancient wisdom of these *auctores*. Instead of prioritising any one interpretation, the *Cosmographia* seems designed to stimulate the kind of deep reading and discrimination of meaning encouraged by the tradition of critical commentary in which Bernard was trained.[134] Such

130 Silverstein, "Fabulous Cosmogony", 116.

131 Gilson, "La cosmogonie", 9: "allégories empruntées aux poètes antiques".

132 *Commentum in Martianum* II.93.109–94.113, ed Westra, 46–47: "Notandum est integumenta equivocationes et multivocationes habere... Hec namque maxime huius voluminis lectores turbant quia, cum aliquid significatum est sub nomine Mercurii, de sermone statim fit transitus ad stellam sub eodem vocabulo. Ideoque in his distinguendis magnam oportet esse lectoris cautelam."

133 William of Conches, *Glosae super Boetium* III, m. 12, ed. Nauta, 203.91–92: "Sed non est curandum de diversitate expositionum, immo gaudendum"; and see Jeauneau, "L'usage de la notion d'*integumentum*", 43–49.

134 Bezner, *Vela Veritatis*, 469, by contrast argues that: "Bernardus' Absicht ist somit nicht... antischolastische Mehrdeutigkeit zu generieren." Whilst I would agree that this is indeed not Bernard's aim throughout the *Cosmographia*, in the case of his treatment of the Incarnation, the Fall and the Trinity (the last of which Bezner, 452–56, interprets somewhat differently from the above discussion) I would maintain that ambiguity (*Mehrdeutigkeit*) is indeed Bernard's deliberate strategy.

characteristics are also evident in Bernard's *Mathematicus*, and in two of the shorter poems, *De paupere ingrato* and *De gemellis*, that have been attributed to 'the school of Tours'. Robert Edwards has described how these narrative treatments of legal and ethical paradoxes do not offer solutions to the cases they discuss, but instead "demand reflection and deliberation on the complexities of experience".[135] Nor was Bernard alone among contemporaries in composing works that demanded high levels of responsiveness from their readers. Abelard's *Sic et non*, written and revised between the 1120s and 1140s, is a collection of thematically organised extracts from patristics, whose apparent contradictions the reader is invited to resolve. In a somewhat similar manner, John of Salisbury's *Policraticus* (*c.* 1159) presents an array of Judaeo-Christian and pagan *exempla* upon which the reader must exercise his powers of discretion in order to determine correct moral action.[136]

The *Cosmographia* belongs in the company of such works. Writing in the aftermath of ecclesiastical censorship of writings that had used Plato and other philosophers to interpret Genesis, Bernard asked his readers to reconsider the role of secular thought in the study of the created world. Rather than suggesting any easy assimilation of philosophy into theology, the *Cosmographia* conveys a sense of the mystery of creation, a subject demanding careful, rational scrutiny from every reader of the work. John of Salisbury described Bernard as "a talented contemporary writer, albeit in the words of the pagans".[137] Although deeply influenced by the poetry and philosophy of antiquity, Bernard's use of the *verba infidelium* is neither proof of his paganism nor simply a classical veneer over Christian doctrine. Rather, his literary treatment of theological doctrine was an invitation to reflect more deeply on the nature of the cosmos and man.

---

135 Edwards, "Poetic Invention", 217. See also Godman, "Ambiguity"; and the expanded version of Godman's article in his *Silent Masters*, 228–93.

136 Von Moos, "Use of *Exempla*".

137 *Policraticus*, ed. Keats-Rohan, 195.115–16: "quidam temporis nostri scriptor egregius, infidelium tamen verbis."

CHAPTER 4

Bernard's Readers

The imitation of ancient authors was an important exercise in the classroom of Bernard of Chartres. But Bernard is said to have enforced a strict distinction between genuine imitation and mere plagiarism, or the sewing on of cloth filched from another source.[1] If Bernard detected such literary theft, he would first reprimand the student in question, but then:

> Sic uero redargutum, si hoc tamen meruerat inepta positio ad exprimendam auctorum imaginem modesta indulgentia conscendere iubebat, faciebatque ut qui maiores imitabatur, fieret posteris imitandus.[2]
>
> Having reproved the student, if an unsuitable theme had invited this theft, Bernard would, with modest indulgence, bid the boy to rise to an exemplary imitation of the authors, and would bring it about that he who was imitating his ancestors would come to deserve imitation by his successors.

The preceding chapters have explained how the *Cosmographia* may be read as a genuine imitation of the ancients—one that did not merely parrot earlier authorities, but adapted and refashioned them to face contemporary philosophical problems. As a genuine imitator, Bernard Silvestris was himself deemed worthy of imitation, and the present chapter is a study of how his *Cosmographia* was studied and imitated—but also challenged—in the course of the twelfth and thirteenth centuries.

The interest of this story lies partly in the light it throws on the wider reorganisation of intellectual life in this period, broadly speaking a transition from the world of the schools to that of the universities. As we have seen, the *Cosmographia* was composed in, and for, a scholarly milieu whose roots stretched back to the great schools of the Carolingian era, but whose emphases

---

1 *Metalogicon* I.24, ed. Hall and Keats-Rohan, 53.76–82: "Quibus autem indicebantur praeexercitamina puerorum in prosis aut poematibus imitandis, poetas aut oratores proponebat et eorum iubebat uestigia imitari, ostendens iuncturas dictionum, et elegantes sermonum clausulas. Siquis autem ad splendorem sui operis alienum pannum assuerat [*cf.* Horace, *Ars poetica* 16], deprehensum redarguebat furtum, sed poenam saepissime non infligebat."

2 *Metalogicon* I.24, ed. Hall and Keats-Rohan, 53.82–85 (repunctuated).

 | DOI 10.1163/9789004268357_006

on the close literary and philosophical study of ancient authors had developed to new levels of sophistication among Bernard's northern French contemporaries. In some ways the *Cosmographia* extended this tradition: in manuscripts the work was equipped with glosses and copied alongside the *auctores*, and it was also imitated—in ways both simple and complex—by some of its readers. On the other hand, already in Bernard's lifetime scholars such as Thierry of Chartres and John of Salisbury were lamenting the decline of the old educational culture, under the pressure of a new vocational breed of student, primarily interested in law and logic, and increasingly concentrated in Paris. Such students allegedly had neither the time nor the finances to pursue the kind of rounded liberal arts education that the appreciation of a work like the *Cosmographia* would require.[3] These concerns about the demotion of fundamental grammatical study continued into the thirteenth century with the foundation of the University of Paris. In Henri d'Andeli's *c.* 1230 *Bataille des VII ars*, for example, "Sir Bernard Silvester, who knew all the languages of the sciences and arts", is one of the *auctores* in the army of Grammar that is defeated in battle by the forces of Parisian Logic.[4] The grammatical *auctores* did not of course disappear entirely from the curriculum: all students still had to learn how to read and write. But the ambitious literary programme of a work like Bernard's did not sit well with an academic culture that was becoming increasingly compartmentalised, as it shifted focus to more pragmatic training in preaching, medicine, and the law.

What follows begins by tracing the reception of the *Cosmographia* through its manuscripts. The first half of the chapter explores the largely unstudied corpus of glosses that accompanies the work in the French and British manuscripts. These glosses are fascinating evidence of early responses to the work and, by virtue of their proximity to the thought-world of the *Cosmographia*, perhaps also of the intentions of Bernard himself. If the glosses are one sign of the *auctoritas* that Bernard enjoyed, so too are the various quotations and imitations of his work surveyed in the second half of the chapter. The most ambitious of these imitations was the *De planctu Naturae* by Alan of Lille, who in a

3 For this theme, see H. Liebeschütz, *Mediaeval Humanism in the Life and Writings of John of Salisbury* (London: Warburg Institute, 1950), 84–90; J. O. Ward, "The Date of the Commentary on Cicero's *De inventione* by Thierry of Chartres (*ca.* 1095–1160?) and the Cornifician Attack on the Liberal Arts", *Viator* 3 (1973): 219–73; Southern, "Schools of Paris"; Ferruolo, *Origins*, 11–26, 140–55; Jaeger, *Envy of Angels*, 278–91; and C. S. Jaeger, "Pessimism in the Twelfth-Century 'Renaissance' ", *Speculum* 78 (2003), 1151–83.

4 *Les "Dits" d'Henri d'Andeli*, ed. A. Corbellari (Paris: Honoré Champion, 2003), 68.329–31: "Et dans Bernardins li sauvages | Qui conoissoit toz les langages | Des scïences et des ars."

sense did to Bernard what Bernard did to his own sources: he wrote a new work whose borrowings from its model served partly to highlight divergences from it. Alan's reservations about the *Cosmographia* were largely theological, and they were shared, as we shall see, by theologians such as Peter Lombard, whose systematic and analytical methods would come to dominate the academic culture of the new universities, leaving little room for the kind of allusive philosophising in which Bernard specialised.

The Standard Gloss

Copies of the *Cosmographia* made during the twelfth and thirteenth centuries survive in twenty-three manuscripts in French and English archives.[5] In many cases, the origins of the surviving manuscripts (where they were copied) and even their medieval provenance (where they were held) are hard to determine. What evidence there is suggests that Bernard was read in a variety of settings on both sides of the Channel. *Ex libris* notes and other indicators of provenance place the *Cosmographia* in the book collections of Benedictines, Cistercians, and canons regular. Intriguingly Bernard's work also appears in

5 For a census of manuscripts, see Appendix II. Medieval library catalogues from France and England record a further nine copies of the work that have not been securely identified with any of the surviving codices. In France, the *Cosmographia* appears in the library catalogue of Richard of Fournival (d. 1260), whose books passed into the collection of the Sorbonne (L. Delisle, *Le cabinet des manuscrits de la Bibliothèque Impériale*, 3 vols. (Paris: Imprimerie Impériale, 1868–1881), II: 531). In England, a catalogue from St Augustine's, Canterbury, compiled between 1375 and 1420, records two copies of the work: one, also containing Abu Ma'shar and the *Asclepius*, is catalogued among astronomical works, the other among the poets (*St Augustine's Abbey, Canterbury*, ed. B. C. Barker-Benfield, 3 vols. (London: British Library, 2008), BA1.1144, BA1.1482). The late-fourteenth-century catalogue of Peterborough Abbey records two books by "Bernardus Silvestris", which may have been the *Cosmographia* (*Peterborough Abbey*, ed. K. Friis-Jensen and J. M. W. Willoughby (London: British Library, 2001), BP21.276, BP21.160.) According to a catalogue of 1355–1360, the well-stocked library of Lanthony Priory, a house of Augustinian canons in Gloucester, possessed two copies of the *Cosmographia* (*The Libraries of the Augustinian Canons*, ed. T. Webber and A. G. Watson (London: British Library, 1998), A16.350, A16.451). Finally, two copies of the work are recorded in Cistercian houses, one at Titchfield in Hampshire, in a catalogue compiled between 1400 and 1404, and one at Rievaulx in a catalogue of *c.* 1190–1200 (*The Libraries of the Cistercians, Gilbertines and Premonstratensians*, ed. D. N. Bell (London: British Library, 1992), P6.31, Z19.127).

volumes that have possible connections to the early universities of Oxford,[6] Cambridge,[7] and Paris.[8] In some cases the *Cosmographia* is found in collections that were bound together at a later date, but in others it appears alongside those works with which it was originally copied. These manuscripts often contain a combination of ancient authorities (Ovid, Macrobius, Martianus Capella, Seneca, Aristotle) and their high medieval successors (Hildebert of Lavardin, Marbod of Rennes, Matthew of Vendôme, Walter of Châtillon, Geoffrey of Vinsauf) that offer a neat indication of the authority that Bernard and some of the other *moderni* enjoyed.[9]

A further sign of Bernard's *auctoritas* in the twelfth and thirteenth centuries is the development of a body of glosses on his *Cosmographia*. As Martine Irvine has written of the practice of glossing:

6 Cambridge, Corpus Christi College 406 [*s*. xiii *in*.] also contains an early English copy of Seneca's *Tragediae*, a descendant of which was used by the Oxford Dominican Nicholas Trevet for his Seneca commentary in 1314–1315. According to R. H. Rouse, "The A Text of Seneca's *Tragedies* in the Thirteenth Century", *Revue de l'histoire des textes* 1 (1971): 93–121 (114), this "would suggest that CCC 406 may have been an Oxford book."

7 Rome, Biblioteca Angelica 401, copied *c*. 1260, contains, in addition to the first statutes of Cambridge University, a number of works by eleventh- and twelfth-century French authors (Marbod of Rennes, Hildebert of Lavardin, Matthew of Vendôme, Peter of Blois, and Bernard's *Cosmographia*), which "help to illustrate the interests of a Cambridge grammarian of the first half of the thirteenth century" (M. B. Hackett, *The Original Statutes of Cambridge University: The Text and Its History* (Cambridge: Cambridge University Press, 1970), 340–43). Hackett also notes (348) that the "mag. J. de Grimesbi" to whom a *quaestio* is attributed on f. 127v in a *Cosmographia* manuscript of otherwise uncertain provenance, Cambridge, Gonville and Caius College 593/453 [*s*. xiii], may have been a Cambridge master.

8 Cambridge, Trinity College O 7 40 contains a text of the *Cosmographia* that is headed by the note "In hoc libello continentur xxv pecie" (f. 71r), which refers to the *pecia* system of copying textbooks adopted by stationers at various universities in the thirteenth century. Only at Paris, however, was this system employed in the Arts faculty. M. R. James, *The Western Manuscripts in the Library of Trinity College, Cambridge: A Descriptive Catalogue*, 4 vols. (Cambridge: Cambridge University Press, 1900–1904), III: 376–79, gives a date of s. xii *ex*. for the *Cosmographia* text, but the *pecia* system appears not to have developed in Paris until the middle of the thirteenth century: see R. H. Rouse and M. A. Rouse, "The Book Trade at the University of Paris, ca. 1250–ca. 1350", in *La production du livre universitaire au Moyen Age: exemplar et pecia*, ed. L. J. Bataillon, B. G. Guyot and R. H. Rouse (Paris: CNRS, 1988), 41–114 (44–47).

9 In a later manuscript a summary of the *Cosmographia* appears in a collection of epitomes of ancients and moderns, including Alan of Lille, John of Hauville, and Peter Riga. See BL Cotton Titus D XX [*s*. xiv/xv], ff. 112v–117v: *incipit*: "Bernardus Silvestris in metris merulam in prosis utique effigans philomenam librum de utroque mundo, maiori videlicet et minori scripsit…"; *explicit*: "… sicque auctor opus consummans secundo libro finem imponit."

> The transmission of glosses and commentary on the same page as the main text is a sign that a text had received institutional validation and had become an object of interpretation and knowledge.[10]

Such annotations had been accruing around the texts of pagan and late antique Christian *auctores* in manuscripts from the eighth century onwards. In the twelfth and thirteenth centuries, however, a number of contemporary works also became objects of academic commentary. This development signalled a new valuation of the contribution of those modern dwarfs, who—in Bernard of Chartres' image—could see further than the giants of antiquity by sitting on their shoulders.[11] Some, but by no means all of the glosses on and *accessus* to Walter of Châtillion's epic *Alexandreis* (1178–1182) have been published.[12] Alan of Lille's *Anticlaudianus* (1181–1184) received an elaborate commentary (*c.* 1212) by Ralph of Longchamps,[13] and other manuscripts of Alan's work contain further evidence of glossing.[14] A third work from the generation following Bernard,

10 Irvine, *Making of Textual Culture*, 373. See further R. Copeland, "Gloss and Commentary", in *The Oxford Handbook of Medieval Latin Literature*, ed. R. J. Hexter and D. Townsend (Oxford: Oxford University Press, 2012), 171–91.

11 This image appears for the first time in the first redaction of William of Conches's *Glosulae de magno Prisciano*, but it is attributed to Bernard of Chartres by John of Salisbury, *Metalogicon* III.4, ed. Hall and Keats-Rohan, 116.46–50: "Dicebat Bernardus Carnotensis nos esse quasi nanos gigantum umeris insidentes, ut possimus plura eis et remotiora uidere, non utique proprii uisus acumine, aut eminentia corporis, sed quia in altum subuehimur et extollimur magnitudine gigantea." See É. Jeauneau, " 'Nani gigantum humeris insidentes': essai d'interprétation de Bernard de Chartres", *Vivarium* 5 (1967): 79–99.

12 Glosses in three manuscripts have been edited as appendixes to Walter of Châtillon, *Alexandreis*, ed. M. L. Colker (Padova: Antenore, 1978), 275–514; part of a freestanding commentary in another manuscript has appeared in *An Epitome of Biblical History: Glosses on Walter of Châtillon's Alexandreis 4.176–274*, ed. D. Townsend (Toronto: PIMS, 2008). Townsend provides an incomplete, but nonetheless substantial, list of annotated *Alexandreis* manuscripts on pp. 20–22 of his edition.

13 See Ralph of Longchamps, *In Anticlaudianum Alani commentum*, ed. J. Sulowski (Wrocław: Zakład Narodowy im. Ossolińskich, 1972), with D. Senekovic, "Der Anticlaudianus-Kommentar des Radulphus de Longo Campo: zur Kommentierungspraxis im Hochmittelalter", in *Sinnvermittlung: Studien zur Geschichte von Exegese und Hermeneutik*, ed. P. Michel and H. Weder, 2 vols. (Zurich: Pano, 2000), I: 475–96.

14 C. Meier, "Die Rezeption des *Anticlaudianus* Alans von Lille in Textkommentierung und Illustration", in *Text und Bild: Aspekte des Zusammenwirkens in Mittelalter und früher Neuzeit*, ed. C. Meier and U. Ruberg (Wiesbaden: Reichert, 1980), 408–549; M. T. Gibson, D. R. Shanzer, and N. F. Palmer, "Manuscripts of Alan of Lille: 'Anticlaudianus' in the British Isles", *Studi medievali*, 3rd ser. 28 (1987): 905–1001.

Joseph of Exeter's *De bello Troiano* (early 1180s) was also an object of medieval commentary.[15] Looking forward to the next generation, there have been recent studies of the glossed manuscripts of Geoffrey of Vinsauf's *Poetria nova* (1208–1213) and the *Graecismus* of Evrard of Béthune (d. *c.* 1212).[16] There is some evidence that other writers from this period—Alexander Neckam (1157–1217) and Andrew Sunesen (*c.* 1161–1228)—added glosses on their own works, perhaps in order to underline the *auctoritas* of their compositions.[17] This, as we shall see, is a possibility that will have to be taken into account with the *Cosmographia* glosses.

The twenty-three manuscripts of the *Cosmographia* that I have studied contain over 550 distinct marginal and interlinear annotations, but only a handful of these have been discussed in previous scholarship, either to shed light on difficult passages or as evidence of the vernacular glossing of Latin texts.[18] The purpose of what follows, then, is to describe the date, contents and purpose of these *Cosmographia* glosses.

A core of around 150 of the glosses appear in most of the manuscripts that contain a complete or near complete text of the work. This standard gloss, as I shall call it, appears to have been added at an early stage in the transmission of the text. It is hard to say exactly when. On palaeographical grounds, a possible twelfth-century date has been proposed for seven of the French and British manuscripts. Only one of these, however, can be dated more securely by internal evidence. BN lat. 15009 is a compilation of three separate manuscripts, bound together at the French abbey of Saint-Victor at the end of the fifteenth century.[19] The *Cosmographia* is found in the third part, which concludes with

---

15 According to Ziolkowski, "Cultures of Authority", 437, n. 76, the commentary in BN lat. 15015, ff. 1r–11v, has been edited by G. B. Riddehough, "The Text of Joseph of Exeter's *Bellum Troianum*" (Doctoral Dissertation: Harvard University, 1951).

16 Woods, *Classroom Commentaries*; A. Grondeux, *Le "Graecismus" d'Évrard de Béthune à travers ses gloses: entre grammaire positive et grammaire spéculative du xiiie au xve siècle* (Turnhout: Brepols, 2000).

17 R. W. Hunt, *The Schools and the Cloister: The Life and Writings of Alexander Nequam (1157–1217)*, ed. M. Gibson (Oxford: Clarendon Press, 1984), 65–66; *Andreae Sunonis Filii Hexaemeron*, ed. Ebbesen and Mortensen, I: 52–54.

18 For difficult passages, see *Cosmographia*, trans. Wetherbee, 150, n. 72, 157, n. 9, 159, n. 30, 160, n. 39, 162, n. 75; and for vernacular glosses, Hunt, *Teaching and Learning Latin*, I: 51–52, 159, who transcribes all the (mostly botanical) vernacular glosses in these manuscripts—fifteen French and two Anglo-Saxon ones in Bodl. Laud misc. 515, and eleven French glosses in Cambridge, Gonville and Caius College 593/453.

19 G. Ouy, *Les manuscrits de l'abbaye de Saint-Victor: catalogue établi sur la base du répertoire de Claude de Grandrue (1514)*, 2 vols. (Turnhout: Brepols, 1999), II: 260–61.

some miscellaneous texts written into the final folios by a known Saint-Victor scribe from the second half of the twelfth century.[20] These texts include a list of French kings and the principal events of their reigns, which runs "usque ad annum VIII regni Philippi [Augusti] regis" (f. 257r). This would give a *terminus ante quem* of 1186/87 for the core 150 or so headings in the manuscript.[21]

An earlier date for the standard gloss may be suggested by the evidence of a twelfth-century philosophical text entitled *De sex rerum principiis*, which purports to be the work of Hermes Trismegistus. One of the work's editors, Mark Delp, has demonstrated that pseudo-Hermes borrowed unacknowledged a number of passages from the *Cosmographia*. However, the *De sex rerum principiis* also incorporates one of the standard glosses, which discusses the fifth element of which Bernard says the heavens are composed:

BN lat. 8808 A, f. 29v (on *Cos*. II.3.10):	*De sex rerum principiis* IX.53–56:
Astronomica Caldeorum est sententia celum et sidera non ex elementis, sed ex quadam quinta natura.[22]	Haec est sententia tam Graecorum quam Latinorum. Et haec Chaldaeorum, qui, diuinius uerum indagare soliti, dicunt caelum et sidera non ex elementis esse, sed ex quadam singulari qualitate.[23]

If Delp is correct in claiming that it is "unlikely that the treatise was composed much after 1150", then this could give an earlier date still for the composition of the annotations.[24] Indeed Nigel Palmer has made the intriguing suggestion

20 F. Gasparri, "Un copiste lettré de l'abbaye de Saint-Victor de Paris au xiie siècle", *Scriptorium* 30 (1976): 232–37.

21 The preceding folios of this part of the manuscript, including the *Cosmographia*, were probably copied in England, given that the texts they contain are either concerned with British history (Nennius's *Historia Brittonum*, a list of British cities, abbreviations of Geoffrey of Monmouth's *Historia regum Britanniae* and Symeon of Durham, extracts from Henry of Huntingdon's *Historia Anglorum*), or were written by authors who spent time in England (Petrus Alfonsi's *Dialogi* and the *Imago mundi* of Honorius Augustodunensis).

22 Here and below I quote from BN lat. 8808 A, which has roughly the same corpus of annotations as BN lat. 15009, but presents a better text.

23 Pseudo-Hermes Trismegistus, *De sex rerum principiis*, ed. P. Lucentini and M. D. Delp, CCCM 142 (Turnhout: Brepols, 2006), 166.

24 *De sex rerum principiis*, ed. Lucentini and Delp, 16.

that the standard gloss may go all the way back to Bernard himself.[25] However, as Palmer himself observes, some of the glosses bear close similarities to the wording of the prefatory summary of the *Cosmographia*, which is probably not by Bernard.[26] It also seems likely that not all these annotations were added at once, but that the standard gloss is a composite of various layers of scholarly activity. Bodl. Laud misc. 515, for example, whose early text led Dronke to choose it as the base manuscript for his edition, contains only 122 of the 150 or so annotations that are common to all the complete or near complete copies of the work. This manuscript—or more precisely its archetype, for the Laudian *Cosmographia* was not copied until the early thirteenth century—may then mark an intermediate stage in the formation of the gloss.

The standard gloss consists mainly of headings that simply paraphrase the accompanying text (e.g. "Equinoctium notat", at *Cos.* II.6.19). These headings would have been useful for medieval readers who often had to navigate texts without the aid of an index or table of contents. In particular the headings that accompany Bernard's long catalogue poem ("Ubi montes famosi", "Arbores fructifere", "Genera natalium", etc.) would have been a great help to the many writers who quoted and imitated this part of the work.[27] In some manuscripts the headings are rubricated in the margins; in others they have crept from the margin into the text itself, making it sometimes hard to distinguish Bernard's words from those of the glossator. Yet if these annotations are to be called headings, it should not be overlooked that a number of them are more complex and interpretative than this label may suggest. Two annotations propose that Bernard made use of *involucra*, those allegorical wrappings that were thought to conceal hidden wisdom in the ancient pagan *auctores*. As in the interpretation of the ancients, these and other broadly allegorising glosses claim that Bernard's descriptions of the gods/planets conceal natural philosophical allusions to the baleful effects of Saturn and the generative powers of the sun-god, Apollo.[28] Again, as with the ancients, these *involucra* could even

25 For some instances of 'self-commentary' in medieval Latin works, see above, 137, and P. G. Schmidt, "The Commentator Knows Better than the Author", *The Journal of Medieval Latin* 18 (2008): 117–29 (124–25), who adds examples from Abbo of Saint-Germain's *De bello Parisiacae urbis*, Jacopo Gaetano Stefaneschi's *Opus metricum*, and Bongiovanni da Cavriana's *Anticerberus*.

26 Palmer, "Plant Names", 42. The preface (*summa operis*) is not found in all the manuscripts and, as Dronke noticed (*Cos.*, 156), it does not exhibit Bernard's characteristic prose rhythm (*cursus*).

27 See the section on "Quotation and Imitation" below.

28 BN lat. 8808 A, f. 30r (II.5.5): "Malignitas Saturni per involucrum significationis exprimitur"; f. 32r (II.5.14): "Pheton innoxius, id est calor temperatus, et veris fructus, id est vernalis fructificatio, de solis proprietate nascuntur."

be read as prefigurations of Christian doctrine. Thus Bernard's description of the Mirror of Providence, which Urania uses to create man, is 'unwrapped' by the standard gloss to reveal the doctrine of divine predestination.[29] And the three rays emanating from Tugaton are held to signify the three persons of the Christian Trinity.[30]

The most interesting headings are those that name ancient authorities as sources for doctrines discussed in the work. The standard gloss contains sixteen such annotations, which I transcribe below, grouped according to *auctor*.[31]

Plato

Distribution of angels (*angeli*) through the cosmos (*Cos.* II.7.2):

Platonicum dogma de locali spirituum positione, de officiis et eorum proprietate.[32]
The Platonic doctrine of the location, duties and nature of the spirits.

Creation of man (*Cos.* II.13.4):

Dicit Plato deum quiddam unde et [MS *quia*] *mundum fecerat ad creationem hominis in cratere reservasse.*[33]
Plato says that God set aside for the creation of man some of that material from which he had also made the universe.

Ensoulment of man (*Cos.* II.4.31; II.10.53–54):

Sententia Platonis animam venientem ad corpora per omne celum sidereum distrahendam, ut de lege siderum vivendi legem emuletur et ordinem.[34]
Plato's teaching was that a soul entering bodies would be drawn through the whole starry firmament, so that it might emulate a law and course of living from the law of the stars.
Platonicum animas ad stellas unde venerunt redire.[35]
The Platonic doctrine that souls return to the stars whence they came.

---

29 *Cos.* II.11.4 (f. 41v): "Describit per involucrum predestinationem divinam."

30 *Cos.* II.5.3 (f. 29v): "Per lucem inaccessibilem supremus genitor, per radios filius et paraclitus tipice demonstrantur."

31 Here again the transcription is from BN lat. 8808 A, with occasional corrections from the ten or so other codices that contain the glosses.

32 35r; *cf.* Calcidius, *Commentarius* CXXVII–CXXXVI, ed. Waszink, 170–77, and J. den Boeft, *Calcidius on Demons* (*Commentarius ch. 127–136*) (Leiden: Brill, 1977).

33 45v; *cf. Timaeus* 41DE, ed. Waszink, 36.

34 28r; *cf. Timaeus* 41E–42A, 47AC, ed. Waszink, 36–37, 44.

35 41r; *cf. Timaeus* 42B, ed. Waszink, 37.

Man's physiology (*Cos.* II.14.127):

Irascentiam Plato dicit in corde esse.[36]

Plato says that anger resides in the heart.

Man's eyesight (*Cos.* II.14.15):

Tria ad visionem necessaria: exterior lux [MS *exteriorem lucem… interiorem*] *aeris, interior, quam mulcebrem ignem Plato, Xenophon diem anime appellat, et corpus cui occurritur.*[37]

Three things are necessary for vision: the external light of the air, the internal light that Plato calls 'fiery light' and Xenophon 'daylight of the soul', and a body that the light meets.

The Chaldeans

The elements (*Cos.* II.13.1; II.3.10):

Dicunt Caldei elementum non dici simpliciter, non solam scilicet substantiam, sed substantiam cum qualitate.[38]

The Chaldeans say that an element is not to be called something simple, that is it is not substance alone, but substance with quality.

Astronomica Caldeorum est sententia celum et sidera non ex elementis, sed ex quadam quinta natura.[39]

It is a doctrine of Chaldean astronomy that the firmament and the planets are not composed of the elements, but of a fifth nature.

Simplicity of the human soul (*Cos.* II.13.15):

Sentiunt Caldei quia anima simplex est, non diversificans in homine spiritus, quemadmodum voluit Iohannicius.[40]

The Chaldeans think that the soul is simple, not a spirit that divides up in man, as Iohannicius maintained.

---

36 51v; *cf.* Apuleius, *De Platone et eius dogmate* I.13, in *De philosophia libri*, ed. Moreschini, 103–4. Bernard himself drew on this chapter of Apuleius to supplement Calcidius's translation of the *Timaeus*, which did not extend to the physiological sections of the dialogue. See above, Chapter 2, n. 65.

37 48v; *cf. Timaeus* 45BD, ed. Waszink, 41–42. I have not been able to trace the reference to Xenophon.

38 44v. I have not been able to trace the references to the Chaldeans in this and the following two glosses.

39 26v.

40 48r; *cf.* Iohannicius, *Isagoge ad Techne Galieni* 17, ed. Maurach, 155: "Spiritus igitur tres sunt: primus, naturalis, sumit principium ab epate; secundus, vitalis, a corde; tertius, i. animalis, a cerebro."

Pythagoras

Principles of created life (*Cos.* II.13.1):

> *Sententia Pitagore duo esse principia: principium sine principio—deum—principium a principio—silvam—eamque diversum cognominat.*[41]
>
> It is a doctrine of Pythagoras that there are two beginnings: a beginning without a beginning—namely God—and a beginning from the beginning— namely *silva*—and that one he names 'the different'.

Homer

The golden chain that binds the universe (*Cos.* II.7.1; II.13.1):

> *Apud Homerum de celo ad terras aurea quedam cathena continuata describitur, que nichil aliud est quam a superis ad infima elementorum firma connexio.*[42]
>
> Homer describes a golden chain stretching from the heavens to Earth, which is nothing other than the stable binding of the elements that runs from the upper to the lower world.
>
> *Cathena Omeri describitur.*[43]
>
> The chain of Homer is being described.

Numenius

Man's body constructed from primordial matter (*Cos.* II.13.3):

> *Numenius philosophus conqueritur deum nec a rerum substantiis nec a stellis nec ab homine silve maliciam penitus extrasisse, quam* [MS *quem*] *eandem maliciam imposturam appellat.*[44]
>
> The philosopher Numenius laments that god did not completely extract from the substance of creation, nor from the stars or from man, the ill-will of *silva* which he calls unavoidable.

Ptolemy

Pantomorphos, the 'painter' of forms in the fixed sphere of the stars (*Cos.* II.3.11):

---

41 44v; *cf. Timaeus* 35A–37C, ed. Waszink, 27–29; Calcidius, *Commentarius* CCXCVI, ed. Waszink, 298.

42 35r; *cf. Illiad* VIII.19, via Macrobius, *Commentarii* I.14.15.

43 44v.

44 45r; *cf.* Calcidius, *Commentarius* CCXCIX, ed. Waszink, 301.

> *Per panthamorphon pictorem in celo Ptolomei sententia innuitur: forme que in inferiori mundo sunt formis superioribus obediunt.*[45]
> The teaching of Ptolemy is alluded to by means of Pantomorphos, the painter in the heavens: the forms in the lower world are subject to the forms in the upper world.

Avienus

Comets produced when Mars is in Scorpio (*Cos.* II.5.11):

> *Sententiam Avieni*[46] *tangit cometas ex planetis bonos ex bonis, malos ex malis certis in temporibus apparere.*[47]
> He touches on the teaching of Avienus that at fixed times good comets appear from good planets, and bad comets from bad planets.

In addition to these *auctor* glosses, there are also two headings in the standard gloss that refer to unnamed philosophers and physicians as authorities on the *genii* assigned to watch over man, and on man's humoural balance.[48]

45 27r. Such a general astrological statement could derive from Ptolemy's *Quadripartitum*, translated by Plato of Tivoli (1136) or from the pseudo-Ptolemaic *Centiloquium*, translated in part by Adelard of Bath (*c.* 1120), and in full by Plato of Tivoli (1138). On these translations, see R. Lemay, "Origin and Success of the *Kitab Thamara* of Abu Ja'far Ahmad ibn Yusuf ibn Ibrahim", *Proceedings of the First International Symposium for the History of Arabic Science*, 2 vols. (Aleppo: Aleppo University, 1978), II: 91–107 (101).

46 This name is corrupt in other manuscripts, e.g. BN lat. 14194, f. 75v (*aveni*) and BN lat. 15009, f. 197r (*anieni*). The name of this fourth-century Latin translator of Aratus had posed problems since antiquity: see A. Cameron, "Avienus or Avienius?", *Zeitschrift für Papyrologie und Epigraphik* 108 (1995): 252–62.

47 31v; *cf.* Servius on *Aen.* X.272 (*Servii Grammatici qui feruntur in Vergilii carmina commentarii*, ed. G. Thilo and H. Hagen, 3 vols. (Leipzig: Teubner, 1881–1887), II: 421: "COMETAE SANGUINEI LUGUBRE RUBENT... Quarum nomina et effectus Avienus, qui iambis scripsit Vergilii fabulas, memorat. Plinius etiam Secundus dicit, cometas stellas esse naturales, quae apparent certis temporibus. Item hoc quoque commemorat, cometas fieri de planetis quinque, unde interdum bonum, interdum pessima significant. Nam si de Venere aut Iove fiant, optima praenuntiant; si de Marte aut Saturno, deteriora." Avienus's iambic version of Vergil does not survive.

48 36v (II.7.7): "Intellexerunt philosophi duos genios, id est angelos, in custodiam hominum a nativitate deputatos"; 46v (II.13.8): "Senserunt phisici hominem in natura sua temperatiorem" (*cf.* Constantine the African, *Pantegni theorica* I.4, in *Opera*, II: 8: "Homo enim omnibus temperatior est animalis speciebus.") Two other glosses providing cross-references to Roman satirists appear in just two of the manuscripts (Cambridge, Corpus Christi College 406, and BL Cotton Cleopatra A XIV): one glosses the whale and dolphin of *Cos.* I.3.419 ("Iuvenalis. Quantum delphinis balena Britannica maior" = *Sat.* X.14), and

The *auctor* glosses reveal a number of things about how the *Cosmographia* was read. It is striking, for instance, that all the annotations cluster around the second book, 'Microcosmus', and that they concern those themes—the relation of the cosmos to man, the nature of *silva*, and the importance of astrology and physiology—that have also been important in my own interpretation of the *Cosmographia*. With one exception, the authorities referred to in these glosses do not appear in the adjacent section of the text.[49] Bernard Silvestris was in general very sparing in references to and quotations from the *auctores*, operating instead—as a good imitator must—on the level of verbal and structural allusion, rather than direct citation. A presumably deliberate consequence of this is that Bernard's myth recaptures some of the archetypal qualities of the ancient creation stories by which he was inspired. One function of these annotations, then, may have been to reassure the *Cosmographia*'s readers of the solid, authoritative foundations upon which the work was built. As the six citations of Plato attest, these foundations were thought to consist largely in doctrines from the *Timaeus*. This was Bernard's primary source. But what is interesting about the other citations is that they all refer to ancients whose ideas appear to have been quarried indirectly from unnamed late antique sources: Apuleius, Macrobius, Calcidius and Servius. These 'secondary' sources tend to feature prominently in modern discussions of the *Cosmographia*, but what was clearly more important to the medieval glossators were the precious philosophical relics they transmitted of lost works by Homer, Pythagoras, and the Chaldeans, and by Numenius, Ptolemy and Avienus.

With its allegorical interpretations and its citation of supporting authorities, the standard gloss confirms the impression that Bernard's early readers viewed—or were encouraged to view—his *Cosmographia* as a work of authority. These glosses would seem to have been the creation of a small number of learned annotators of the text. They do not, however, appear to be prepared for school use, for in school manuscripts one would expect to find a wider range of notes on vocabulary, rhetoric, grammar and basic interpretation.[50] Only one

---

one the roach, barbel and plaice of I.3.438 ("Aliter pecten. Oratius: pectinibus patulis iactat se molle Tarentum" = *Sat*. II.4.34). The second gloss is mysterious, for *pecten* (scallop) is obviously not synonymous with any of these fish, but nor is it a metrically viable variant reading.

49 The exception is the heading citing Homer in the gloss on Bernard's words at *Cos*. II.7.1: "In lunari enim limite, ubi auree Homeri quasi medietas est cathene..."

50 For a useful set of criteria, see G. A. Wieland, "The Glossed Manuscript: Classbook or Library Book?", *Anglo-Saxon England* 14 (1985): 153–73.

copy of the *Cosmographia* is furnished with such glosses, and it is to this manuscript that we should now turn.

The Laudian Gloss

The most remarkable of the French and English manuscripts, Bodl. Laud misc. 515, was copied in the first quarter of the thirteenth century.[51] In addition to 122 of the standard glosses, which appear as rubricated headings, this manuscript contains more than 250 further glosses written in a single, roughly contemporary hand. Almost all these additions are apparently unique to the Laudian manuscript, but the fact that a few of them are substantially identical to the standard glosses in other manuscripts indicates that the glossator certainly copied some (and possibly many) of his notes from another manuscript—one that is now lost or has not been examined for the present study.

Whereas we know little about the provenance of most of the British and French manuscripts, this book can be quite convincingly tied to Waltham Abbey, a house of Augustinian canons in Essex.[52] For its short history, Waltham had a distinguished intellectual past. The community was established in the 1030s by the Danish thegn Tovi the Proud, and then endowed by Harold Godwineson (*c.* 1060) with a further eleven canons. One of these was the German scholar Adelard of Liège, who was one of a number of Lotharingian clerics to have entered the English Church in the eleventh century, some of whom brought with them interests in Arabic science.[53] In the 1120s, Adelard's

---

51 H. O. Coxe, *Bodleian Library, Quarto Catalogues II: Laudian Manuscripts. Reprinted from the Edition of 1858–1885, with Corrections and Additions, and a Historical Introduction*, ed. R. W. Hunt (Oxford: Bodleian Library, 1973), 372–74; and *Cos.*, 66 (Dronke had this date "confirmed... palaeographically by R. W. Hunt.")

52 A distinctive fourteenth-century pressmark on f. 1v ("lxxxxviii al. ca." [= almarium canonicorum]) and a thirteenth-century *ex libris* with anathema clause on f. 2r ("Hic liber est de armario sancte Crucis de Waltham, quem qui abstulerit vel furatus fuerit vel hunc titulum scienter deleverit et mali<ci>ose excommunicatum se esse noverit") indicate that the manuscript was owned by Waltham Abbey.

53 For Adelard's appointment, see *The Waltham Chronicle: An Account of the Discovery of our Holy Cross at Montacute and its Conveyance to Waltham*, ed. L. Watkiss and M. Chibnall (Oxford: Clarendon Press, 1994), 28–29. The Watham Chronicle was written shortly after the community's refoundation as a house of Augustinian canons in 1177. Lotharingians who brought with them native interests in Arabic science were Robert, Bishop of Hereford (1079–1095) and Walcher of Great Malvern (d. 1135): see C. Burnett, *The Introduction of Arabic Learning into England* (London: British Library, 1997), 14–16.

son, Peter, taught Latin and verse composition at the school, and it would have been around this time that Lawrence of Durham (*c.* 1100–1154), a native of Waltham and the author of, among other things, a prose and verse *Consolatio de morte amici*, studied at the school.[54]

From *c.* 1200, there survives a catalogue of 132 books that belonged to the abbey. These were perhaps a personal bequest of one of the canons rather than a list of the abbey's core collection.[55] Among them are 62 liberal arts texts, which contain a remarkable spectrum of authors relevant to the thought-world of Bernard Silvestris—Plato, Macrobius, Martianus Capella, Boethius, Adelard of Bath, Thierry of Chartres, and the *Declamationes* of pseudo-Quintilian.[56] This list is indicative of the wide intellectual interests of the Augustinian order in England at this time, which are also evident in the extensive (if idealised) reading list in the liberal arts that Alexander Neckam included in his *Sacerdos ad altare*, written in the second decade of the thirteenth century for students at the Augustinian abbey of St Mary's, Cirencester.[57] Laud misc. 515 does not appear in the Waltham booklist, but the manuscript offers further evidence of the flourishing intellectual life at Waltham in this period.

Given the likelihood that at least some of the abbey's books came from France, is it possible that the *Cosmographia* gloss was also composed there before crossing the Channel to Essex? The internal evidence is mixed. At some points the gloss demonstrates a familiarity with French matters. The glossator remarks, for example, that the sorb-apple grows in the region of Montpellier.[58]

54 *Waltham Chronicle*, ed. Watkiss and Chibnall, 66–67, xxvi.

55 *Libraries of the Augustinian Canons*, ed. Webber and Watson, A38, esp. 428: "The small number of patristic texts in the list suggests that this may not have been the community's main book collection ... It is possible that the list may represent, at least in part, a collection formed by an individual scholar."

56 R. M. Thomson, *Books and Learning in Twelfth-Century England: The Ending of 'Alter Orbis'* (Walkern, Herts.: Red Gull, 2006), 94, remarks that the list includes "unusual twelfth-century writings such as Thierry of Chartres' commentary on Cicero's *De Inventione*, and grammatical glosses by William of Conches and Peter Helias ... Serious teaching of the liberal arts was being undertaken at Waltham, and not just at a low level. And someone had been to Paris." The editors of Adelard of Bath, *Conversations with his Nephew*, ed. Burnett *et al.*, xviii–xix, suggest that Waltham may have played an important role in the transmission of Adelard's writings.

57 Alexander Neckam, *Sacerdos ad altare* VIII–XVIII, ed. C. J. McDonough, CCCM 227 (Turnhout: Brepols, 2010), 174–214. See also D. Postles, "The Learning of the Austin Canons: The Case of Oseney Abbey", *Nottingham Medieval Studies* 29 (1985), 32–43.

58 *Cos.* I.3.291 (f. 192v): "SORBA circa patriam Montis Pessulani herba est valde constringentis nature, et est acidissima in viriditate." On the other hand, in Bernard's forest catalogue (*Cos.* I.3.353), the glossator emends the proper noun *vastem* (i.e. the Gâtine) to the adjec-

Elsewhere, he mentions that the *Cosmographia* was recited to Pope Eugenius III when he was in France, an event apparently unattested in any other source.[59] However, such knowledge does not necessarily indicate a French origin for the gloss, as opposed to the work of a scholar who had travelled in France or was otherwise familiar with the country. This second possibility looks the more likely when the manuscript's vernacular glosses are taken into consideration. For although eighteen of these glosses, which are clustered around Bernard's plant catalogue, are in French, two are in English. Such proportions are common in English manuscripts of Latin texts from this period. They indicate, as Tony Hunt has written, that users of these manuscripts "had a higher level of culture and may not have had much knowledge of English at all."[60]

The case for an English, and particularly a Waltham origin for the gloss is strengthened by the impression (documented below) that the Laudian annotator was using books listed in the abbey's catalogue. This raises the question of whether traces of the glossator's work may be found in other surviving manuscripts from Waltham. Twenty-seven such codices have been identified. I have examined all but five of these, and while a couple contain quite extensive glossing, sadly none looks to have been annotated by the intriguing scholar who worked on Laud misc. 515.[61]

The Laudian glossator was writing for readers with quite a high level of Latin proficiency. This may seem to go without saying, but challenging texts were not always the exclusive preserve of the most advanced students. Oseney Abbey in Oxfordshire, for example, another house of Augustinian canons, possessed in the thirteenth century a copy of Calcidius's *Timaeus* that contains many simple lexical glosses, which suggest that even difficult works could be made to serve a wide constituency of readers.[62] By contrast, although the Laudian *Cosmographia* does contain a few synonym glosses, these tend to be relatively sophisticated, for example "VERTIGINE: revolutione" (*Cos.* I.1.49; f. 183v), and "COLOXUM, id est habitacio, de colo, -lis" (II.14.119; f. 218r). A good proportion

---

tive *vastam*—to modify *silvam* in the following line—but this may perhaps have been an alternative name for the wood.

59 For discussion of this gloss, see above, Chapter 3, n. 7, and below, n. 93.

60 Hunt, *Teaching and Learning Latin*, I: 13; and see above, n. 18.

61 See N. R. Ker, *Medieval Libraries of Great Britain: A List of Surviving Books*, 2nd ed. (London: Royal Historical Society, 1964), 192–93. I have not examined the manuscripts held in private collections, nor the Magdalene (Cambridge), Princeton or Görlitz manuscripts. There is extensive glossing in Bodl. Auct. D 4 22 (Gospel of Matthew) and Bodl. Laud lat. 109, ff. 114r–121r (*Pictor in carmine*).

62 Just the first page of the dialogue (Bodl. Digby 23, f. 4v) contains these examples: "CERTE: sine ambiguitate", and "PORRO: sed".

of these lexical glosses explain the various Greek terms that pepper Bernard's text—words such as *pyr*,[63] *gerarchia*,[64] *Helix*,[65] and *aplanon*.[66] For help with some of these, the Laudian annotator may have turned to the recent and popular *Graecismus* by Evrard of Béthune.[67] Elsewhere too the glossator tends to focus on rather advanced literary topics, such as the identification of recondite verse forms—perhaps with the aid of "Seruius de metris", item 81 in the Waltham booklist—and the labelling of rhetorical figures.[68] In this connection, it may be noted that the *Cosmographia* is preceded in Laud misc. 515 by Geoffrey of Vinsauf's *Poetria nova* (apparently copied by the same scribe) and by a number of metrical saints' lives that may be school poetry.[69]

Much of the glossator's attention, however, was focused on furnishing the catalogue poem with an apparatus of encyclopaedic information to explain Bernard's references to an array of celestial bodies, mountains, rivers, animals, plants and mythological/historical figures. Such notes make up around half his

---

63 *Cos.* I.2.15.12 (f. 187r): "Pyr grece, ignis latine. Inde PIROSIS, quod idem est quod incendium."

64 *Cos.* I.3.30 (f. 188r): "GERARCHIA dicitur a 'gera', quod est sacrum, et 'archos', quod est princeps. Inde gerarchia principalis sanctitas, vel principatus sanctitatis."

65 *Cos.* I.3.78 (f. 188v): "HELICE grece, ursa latine."

66 *Cos.* II.3.9.8 (f. 200v): "APLANON, ab 'a' quod est sine, et 'planes' quod est error, id est sine errore. Unde planete grece, erratici latine."

67 *Cos.* I.3.80 (f. 189r): "LIRA duas habet significationes: sulcus et cithara. Unde: 'Pollice tango liram, sed tango vomere liram' "; *cf.* Evrard of Béthune, *Graecismus* X.146–48, ed. J. Wrobel (Breslau: Koebner, 1887), 81: "Est lyra qua cantas, est lira cauernula iuxta | Culturam sic a liro pro deuio dicta: | Pollice tango lyram, facio cum uomere liram." See also nn. 81 and 92, below. Another mnemonic verse for *taxus* (masc. = badger; fem. = yew-tree) occurs at I.3.276 (f. 192r): "Hic TAXUS gallice tessun, hec taxus arbor. Unde versus: 'Hic taxus pecus est, hec taxus dicitur arbor, | Hoc taxum truncus dumtaxat dicitur inde.' "

68 Verse forms: *Cos.* II.4 (f. 201r): "Secundum Servium metrum hoc dactilicum alcmanium tetrametrum acatalecticum" (*cf.* Servius, *De centum metris*, in *Grammatici Latini*, ed. Keil, IV: 456–67 (460): "De alcmanio. Alcmanium constat tetrametro acatalecto, ut est hoc, 'virgo decens comulas religa tibi' "); II.12 (f. 213r): "Metrum dactilicum ymenaicum dimetrum ypercatalecticum."

Rhetorical figures: *Apharesis*: *Cos.* II.1.2.2 (f. 198r): "MISSUM, id est omissum: afferesis"; II.14.115 (f. 218r): "COLUMEN. Afferesis est, id est incolumen"; *Hendiadys*: *Cos.* II.9.5.5–6 (f. 209v): "GIRIS, AMFRACTIBUS: endiadis"; *Antonomasia*: *Cos.* II.3.3.8 (f. 199v): "Uranon, grece palacium; dicitur etiam uranon celum, quia per antonomasiam Urania celestis musa"; *Anthropospatos*: *Cos.* I.1.31 (f. 183r): "AUCTOREM TERRERE. Protospatos [*sic*] est, id est humana passio. Non enim terror cadit in deum silve. *Domine, ne in furore tuo* [Ps. 6:1], et cetera."

69 For the connections between metrical saints' lives and the schools, see Dolbeau, "Un domaine négligé".

glosses in the manuscript. To give just a sample of these, the glossator elucidates two cryptic couplets on the star signs Capricorn and Cancer;[70] he claims that the *devia claustra* of the mountain catalogue are the Pyrenees, which 'deviate' from the Alps mentioned in the previous line;[71] he notes that the Alpheus and Arethusa, though named separately by Bernard, are in fact the same river;[72] a line on the partridge (*perdix*)—"would that he had studied less and lived longer!"—is explained as an allusion to Perdix, who was pushed by his jealous uncle Dedalus from a high tower into the sea, but then transformed by Athena into the bird named after him;[73] and mention of the herb rue (*ruta*) calls forth a grim anecdote about King Mithridates' use of the plant.[74] The glossator's most useful source for this kind of information was undoubtedly Isidore of Seville's *Etymologiae*, which figures as item 75 on the Waltham booklist. At ten citations, Isidore is the most frequently mentioned author in the glosses, which also include much unacknowledged use of the *Etymologiae*.[75] By contrast, there is no evidence that the glossator knew the most important thirteenth-century philosophical encyclopaedia, the *Speculum naturale* of Vincent of Beauvais (*c.* 1190–1264?), which contains Aristotelian philosophy entirely absent from the Laudian gloss.

As in the standard gloss, the Laudian glossator also cites a number of other ancient authors. Often these are classical poets, whose verse is quoted alongside particular lines in the catalogue poem. Such notes resemble the *apparatus*

---

70 *Cos.* I.3.125–28 (f. 189v): "De Capricorno ponit quatuor versus et de eius opposito, scilicet Cancro. In primis duobus ostendit eorum oppositionem, in secundis eorum naturas, quia secundum accessum et recessum ad illa duo proveniunt diversitates tocius anni."

71 *Cos.* I.3.198 (f. 190v): "CLAUSTRA DEVIA, id est Pireneus mons Hispanie, Haspremunt gallice. Mons ille deviat ab Alpibus." *Devia claustra* may rather stand in apposition to *Alpes* (l. 197) as an example of rhetorical *interpretatio*, where one idea is repeated in synonymous terms—so "pathless defiles".

72 *Cos.* I.3.251 (f. 191v): "ALPHEUS et ARETHUSA idem est, et nascitur in Archadia, et ibi dicitur Alpheus. Ibi cadit in terram et surgit sub nomine Arethuse, primo in Orthigia, que Delos dicitur, secundo in Sicilia."

73 *Cos.* I.3.467 (f. 195r): "Nota. Dedalus precipitavit PERDICEM, nepotem suum, in mare, eo quod invenit eum sagacem, sed miserata Pallas mutavit eum in avem"; *cf.* Ovid, *Met.* VIII.236–59.

74 *Cos.* I.3.399 (f. 194r): "Nota. Metridates rex Ponti usus est frequenter ruta et sale propter venenum. Postea devictus in bello hausit venenum et inde mori non potuit, sed cum interius nimis vexaretur, caput suum fecit amputari."

75 Isidore is quoted in glosses on I.3.138 (f. 189v), 186 (f. 190v), 193 (f. 190v), 195 (f. 190v), 221 (f. 191r), 270 (f. 192r), II.13.2.2 (f. 214r), and II.14.51 (f. 217r). Unacknowledged borrowings occur, for example, at *Cos.* I.1.42 (f. 183v): "ALUMPNUS est qui alit et alitur" [*cf. Etym.* X.3], and I.3.471 (f. 195r): "VULTUR a tardo volatu, sicut milvus a molli volatu" [*cf. Etym.* XII.7.12, 58].

fontium of a modern edition. The most frequently mentioned poet is Ovid. A couplet on the constellations of the Crow, the Serpent, and the Crab is taken to be a learned variation on an Ovidian pentameter;[76] the rare epithet *Sigeus* (Trojan) and the grove of Aricina are linked to likely sources in the *Metamorphoses*;[77] and alongside Bernard's mention of the woodpecker (*picus*) the glossator quotes lines from Ovid that describe but do not name the bird.[78] Besides Ovid, the gloss notes reminiscences of other ancients poets in lines on the curvature of the heavens (Lucan), the gazelle's antlers (Martial), and the lofty winter oak (Vergil).[79]

In addition to poets, the gloss draws on a range of philosophical prose. As in the standard gloss, Plato is the most frequently mentioned author, albeit on only two occasions. At one point, the glossator cites Plato to elucidate Bernard's idea of an ever-repeating cycle of creation, whereby forms proceed from first principles, through genus, species and individual, and back to first principles;[80] at another, Plato is used to add technical precision to a discussion of the moon's

76 *Cos.* I.3.115–16: "Cum Corvo loca nota tenet distentior Ydrus, | Et Crater, Cancri sub regione situs", with the gloss (f. 189v): "Ovidius: 'Anguis, Avis, Cancer, sidera iuncta micant'" [= *Fast.* II.266].

77 *Cos.* I.3.249 (f. 191v): "SEGEIA, a Sigeo promontorio. Unde Ovidius: 'Dextera Sigei, Rethei leva profundi'" [= *Met.* XI.197]; *Cos.* I.3.341 (f. 193r): "Hec ARICINE, -nes, vel Aricina, -ne, silva iuxta Romam, et etiam vallis. Unde Ovidius de Numa mortuo: '... nam coniunx urbe relicta | Vallis Aricine densis latet abdita silvis'" [= *Met.* XV.487–88].

78 *Cos.* I.3.477: "Litoris alcinoe [*sic*], nemoris custodia picus", with the gloss (f. 195v): "Ovidius de PICO: 'Et iratus diro fera robora rostro | Figit et iratus longis dat vulnera ramis'" [*cf. Met.* XIV.391–92: "Indignatus avem duro fera robora rostro | Figit et iratus longis dat vulnera ramis."].

79 *Cos.* I.3.137–38: "... septena planetas | Sidera devexum currere mandat iter", with the gloss (f. 189v): "Ysidorus: 'Convexa sunt extrema celi a curvitate dicta', devexa inferiora. Unde Lucanus: 'Mundi devexa secutus'" [= Luc. X.39; *cf. Etym.* XII.5.6]; *Cos.* I.3.206: "... et in bubalo cornua frontis honor", with the gloss (f. 191r): "Marcialis: 'Dente timetur aper, defendunt cornua c<ervum>: | Inbelles damme quid nisi preda sumus?'" [= Mart. XIII.94] (angled brackets here and below contain my conjectures of text lost to marginal trimming); *Cos.* I.3.295: "Esculus alta solo ...", with the gloss (f. 192v): "In Georgicis legitur quod radices esculi tantum extendunt in terra quantum frondes supra" [*cf. G.* II.291–92: "Aesculus in primis, quae quantum uertice ad auras | Aetherias, tantum radice in Tartara tendit."] See also n. 48, above.

80 *Cos.* I.4.3 (f. 196r): "Plato dixit genera et species esse ydeas, et nascente quolibet homine exiit quedam ydea ab hac generali ydea—animal—sive ab hac speciali—homo—que efficiebat solum esse solum, et sic de quolibet. Ipso vero mortuo rediit illa ydea individualis ad sua priora. Et hanc opinionem tangit hic auctor"; *cf.* Calcidius, *Commentarius* CCCXXXIX, ed. Waszink, 32, with the idea of celestial return from *Timaeus* 24B.

phases.[81] Neither of these citations appears in the *Timaeus*, but the Platonic attribution might be explained by references to these opinions in a gloss on that dialogue, or else by a marginal reference to Plato in a manuscript of one of the authors from whom the opinions in fact derive. Items 91 and 92 in the Waltham Abbey booklist, "Macrobius et Plato simul" and "Glose super eos"—both lost, or as yet unidentified—may be likely candidates. Besides Plato, there are further citations of Pliny, in connection with the perfectly round form of the cosmos, and of "Marcianus in astrologia" (perhaps from item 94 in the Waltham booklist: "Astrologia Marciani et glose"), on the nine circles of the firmament.[82]

Most interestingly, the Laudian glossator also cites two authorities from the new medical literature by which Bernard was himself inspired—and which it would have been only natural for an interpreter of his work to consult. One gloss is a learned digression on a verse that reports Cato's esteem for cabbage. Cato's *De agri cultura*, where cabbage is praised (ch. 156), was not at all widely known in the Middle Ages; instead, the glossator offers a summary of the chapter on Catonian and other cabbage lore in Dioscorides' *De materia medica*, a first-century AD pharmaceutical treatise quoted by the glossator from an eleventh or twelfth-century revision (possibly by Constantine the African) of the sixth-century Latin translation. Cabbage is presented here as an effective treatment for a range of maladies, and particularly beneficial for young boys who wash in the urine of those who have consumed the plant.[83] The second gloss

81 *Cos.* II.1.4.2 (f. 198v): "Nota. Opinio Platonis fuit lunam non habere lumen nisi ex sole. Qui dicit quod post remotionem unius diei dicitur monoydes, quando apparet media dicitur dicotomos vel diatomos, quando fere plena amphicirtos, quando plena pansirenos quasi tota serena, vel pansilenos quasi plenilunium." For the illumination of the moon, see Macrobius, *Commentarii* I.19.8; and for the phases, Martianus Capella, *De nuptiis* VII.738; Macrobius, *Commentarii* I.6.54–56; and Evrard of Béthune, *Graecismus* VIII.68–71.

82 *Cos.* I.4.9.7 (f. 197r): "Nota. Quodlibet celum naturaliter rotundum est et spericum. Unde licet aqua se extenda<t> longe et late, tamen secundum Plinium naturaliter sperica est, ut patet in gutta que pen<det> in fronde, vel illa que cadit in pulvere<m>"; *cf.* Pliny, *Historia naturalis* II.163. *Cos.* I.3.61 (f. 188v): "In hoc loco distinguitur firmamentum in quinque regiones, scilicet in quinque paralellos et in duos coluros, et in zodiacum et in galaxiam, et infra de novem circulis facit mencionem, et istos novem circulos assignat Marcianus in astrologia, et preterea decimum, scilicet orizonta"; *cf.* Martianus Capella, *De nuptiis* VIII.817–26.

83 *Cos.* I.3.399 (f. 194r): "Diascorides in libro suo ait: Brassica desiccative virtutis est, plagas glutinat, sanat inflammaciones iam induratas, id est rancles. Eripsilas induratas, id est sanctum ignem, et herpetas curat, id est dertres gallice. Cato dicit brassicam canceratos curare, si in aqua calida pota sit. Cruda contrita bis in die pota inpetigines sine ulceratione sanat, id est guttaiwere gallice. Fistulas elixata decutit, id est festre gallice, et

contains another medical digression—this time drawing on Isaac Iudaeus's *De dietis*, again perhaps translated by Constantine the African—on the apparent problem of why Bernard described man's lungs as soft and airy.[84] Like a number of the works discussed above, Isaac's treatise also appears (as item 102) on the Waltham Abbey booklist.

Besides such *auctor* glosses, the Laudian manuscript contains a number of annotations that creatively interpret Bernard's allegory. Towards the beginning of the work, for example, Bernard describes how Noys bound together soul and matter with a kind of glue (*quodam glutino*): this provokes a long note on the macrocosm/microcosm analogy that gives further proof of the glossator's interests in physiology. The extremes of fire and earth, he writes, are bound together by the intermediary elements of air and water, just as body and soul are bound together by spirit and complexion, a binding that can be fatally undone by the desiccating effects of hellebore.[85] Elsewhere the glossator

---

tumores spargit. Urina eius qui eam comederit nervis in fomento calido est utilis, id est in balneo fomentum est in quocumqe liquore. Pueri etiam si ea frequenter laventur numquam erunt debiles. Hec Cato. Lege ergo litteram: 'Brassis que [*sic*] probata Catoni', id est experimentata, vel probata, id est commendata, quia Cato maxime commendat, ut narrat Diascorides. Narrat enim quod per multos annos non viguit alia phisica Rome, sed electuario inde composito multos curabat Cato." This gloss summarises, with the addition of French glosses, parts of *De materia medica*, s. v. *brassica*. Unable to consult either of the early modern editions of *De materia medica*, I have compared this gloss with New York, Pierpont Morgan Library M 760, ff. 15v–16r. For the tradition of *De materia medica*, see J. M. Riddle, "Dioscorides", in CTC, IV: 1–143.

84 *Cos*. II.14.123–26 (f. 218r): "Ysaac in dietis: PULMO flegmatico pertinet humor<i> [*cf.* Isaac Iudaeus, *Liber dietarum particularium*, in *Omnia Opera Ysaac* (Lyons: Barthélemy Trot, 1515), ff. 103r–156r (138v): "Pulmo…ceteris membris minus nutrit generans humorem phlegmati pertinentem"]. Sed quid est quod auctor dicit eum MOLLEM ET AERIUM? Solutio: <cor> ferventissime nature est et de facili posset abuti ca<lore> suo, et seipsum dissolvere nisi temperaretur. Sed cum pulm<o> sit aquatice nature, substantialis frigiditas aque cum accidentali humiditate non sufficeret cordis fervorem temperare. Extenditur ergo pulmo sursum in duas ligulas, et conprehendit cor a dextra et a sinistra. Ante autem et retro cor ascendit a pulmone quedam pinguedo, et conprehendit cor; sed omnis pinguedo calida et humida est, unde nature aerie est, id est substantialiter humida. Concurrunt ergo substantialis humiditas pinguedinis et substantialis frigiditas pulmonis ad temperandum fervorem cordis. Et attribuit auctor illam pinguedinem que conprehendit cor ante et retro pulmoni."

85 *Cos*. I.2.15 (f. 187r): "Sicut ignis et terra inter se dissonancia duobus mediis, aere et aqua, connectuntur, ita corpus et anima duobus, scilicet humore et spiritu. Humores enim magis consonant corpori, utpote cola, melancolia, flegma, sanguis. Spiritus vero grossior est anima, et accedit ad naturam anime. Inter humores vero et spiritum quedam est consonancia, sicut inter aquam et aera. Sicut enim aqua de facili resolvitur in aera et nutrimentum est aeris, eodem modo humores nutrimenta sunt spiritus. Unde dicitur in

suggests an interpretation of the three instruments that Noys gives her daughters for the creation of man: the Mirror of Providence is theology, the Tablet of Destiny astronomy, and the Book of Memory physics.[86] As in the standard gloss, these interpretations sometimes take a theological turn. When Noys is about to give her daughters their creative tools, Bernard writes of the "multiplicity, seriousness and weight" of the challenge ahead: the glossator makes the slightly strained suggestion that this is a biblical allusion to Ws. 11:21, where God is said to have ordered all things by measure, number and weight.[87] Elsewhere, and again like the standard gloss, the Laudian annotator reads the three rays emanating from Tugaton as an allegory of the Trinity.[88] But he takes this trinitarian reading further, arguing that Bernard's mention of *notiones* in the divine mind (Noys) refers to five distinct 'ideas' that theologians have predicated of the Trinity: *paternitas* (fatherhood), *filiatio* (sonship), *processio* (ability to proceed), *spiratio* (ability to breathe forth), and *innascibilitas* (inability to be born).[89]

---

phisica: 'Spasmus ex elleboro, mortale' [Hippocrates, *Liber aphorismorum* V.1, in *Articella* (Venice: Baptista de Tortis 1487), ff. 9r–45v (31r); the *Aphorismata* is item 101b in the Waltham booklist]. Spasmus est contractio nervorum, gallice craupe [*sic*]. Et hec est ratio quare mortale: elleborum acutissima est herba; cum igitur ex purgacione facta per elleborum contingit spasmus, hoc contingit ex nimia inanicione humorum. Cum ergo humores nutrimenta sint spiritus, deficientibus humoribus deficiet spiritus. Sed spiritus est conglutinatio corporis et anime; ergo deficiente vinculo compositum dissolvetur. Ex his patet quod bene dici potest complexionem ex quatuor humoribus surgere, et complexio et spiritus sunt vinculum extremorum. Sed notandum quod hoc dictum est in phisica de spiritu vitali. Est enim triplex spiritus: vitalis, animalis, naturalis [*cf.* Iohannicius, *Isagoge*, quoted above, n. 40, and item 101a in the Waltham booklist]. Vitalis animam et corpus conglutinat; animalis maxime vigens in cerebro et subservit memorie, rationi et imaginacioni, et preter nobilem operacionem quam facit in cerebro facit etiam sensum et motum; naturalis viget in epate et deservit virtuti appetetive, retentive, digestive et expulsive."

86 *Cos.* II.11.3 (f. 211v): "Per speculum providencie intellige theologiam, per tabulam fati astronomiam, per librum recordacionis phisicam."

87 *Cos.* II.11.3.1–2: "Verumtamen si, pre rerum multitudine, auctoritate et pondere…", with the gloss (f. 211v): "Alludit illi quod dicitur: 'Deus creavit omnia numero, pondere et mensura.'"

88 *Cos.* II.5.3.3–8 (f. 202v, with glosses in brackets): "Ea igitur lux inaccessibilis intendentis reverberat oculos, aciem preconfundit, ut, quia lumen se [id est filius] defendit a lumine [sicut sol se defendit a nostro lumine, sancta trinitas a lumine intellectus], splendorem ex se videas caliginem peperisse. Ex splendore igitur vel infinibili vel eterno alter se radius exerebat, ut ex primo secundoque suboriretur et tercius [id est spiritus sanctus]."

89 *Cos.* I.2.8.4 (f. 185v): "INTERNAS [*Dronke* ETERNAS] NOSCIONES. In theologia quidam dicunt tres esse nosciones trium personarum trinitate: paternitatem, qua hec persona, pater, distinguitur ab hac persona, filius; filiacionem, qua hec persona, filius, distinguitur a spiritu sancto; processionem, qua spiritus sanctus distinguitur ab aliis personis. Et

The Laudian gloss suggests that the *Cosmographia* was deemed a work of sufficient authority to merit the pains taken by its annotator to explicate allusions, provide cross-references to ancient authors, and offer elaborate philosophical and theological interpretations. On a couple of occasions the glossator does for his *auctor* what William of Conches did for Boethius—saving Bernard's authority against charges (raised by students?) of error and inconsistency.[90] But he is not always a slave to the text. Twice he charges Bernard with abuses of diction. At one point the author is criticised for writing that Physis divided between man's nostrils the 'scent-emitting', rather than the 'scenting' power of smell.[91] In another passage the glossator chastises Bernard—not entirely justifiably—for describing Silva with the Greek loanword *usia*, which denotes matter together with form, whereas, according to our commentator, Silva is in fact matter *without* form, and therefore an *ypostasis*.[92]

---

dicunt quod quelibet noscio est ipsum cuius est noscio: unde paternitas est pater, filiacio filius, processio spiritus sanctus. Nichil enim est in deo quod non sit deus. Alii dicunt quatuor esse nosciones, scilicet tres predictas et quartam, scilicet spiracionem, que comunis est patri et filio; ea enim distinguitur utraque persona a spiritu sancto. Alii addunt quintam, scilicet innascibilitatem, que tamen est noscio patris. Et hac ratione dicunt quinque esse nosciones, quia omnis consonancia musica est in deo, sed quinque sunt consonancie in musica: diatessaron, diapente, diapason, kaydiapente, kaydiapason. Similiter quinque nosciones in trinitate." This gloss combines the patristic doctrine of *notiones*—summarised, for example, in Stephen Langton's *Quaestiones*, ed. S. Ebbesen and L. B. Mortensen, "A Partial Edition of Stephen Langton's Summa and Quaestiones with Parallels from Andrew Sunesen's Hexaemeron", *Cahiers de l'Institut du moyen-âge grec et latin* 49 (1985): 25–224 (208): "Quinque sunt notiones: tres in patre (paternitas, innascibilitas, spiratio), duae in filio (filiatio et spiratio), una in spiritu sancto (processio)"—and the musical intervals discussed in Boethius, *De institutione musica* I.16; Macrobius, *Commentarii* I.6.43–44, II.1.14–20; and Calcidius, *Commentarius* XLVI.

90 *Cos.* I.3.46–47: "... et geometra Thales; | In stellis lepidum dictat Maro, Milo figurat", with the gloss (f. 188r): "Thales arismeticus fuit, et Milo geometer, sed quia geometria preparat viam in arismeticam, auctor attribuit uni quod fuit alterius"; *Cos.* II.1.9–10: "Ecce mundus cui Noys vita ...", with the gloss (f. 198r): "Alibi dicit quod Endelichia est vita mundi [*cf.* I.2.13.17: "vita ... rerum"], sed quia Endelichia vitam trahit a Noi, ideo Nois vocatur hic vita mundi." For William of Conches, see above, Introduction, 7–8.

91 *Cos.* II.14.99–100: "Inter ... olentem naris odorem | Dividit", with the gloss (f. 218r): "OLENTEM, id est a<ro>maticum. Abusio in dictione."

92 *Cos.* I.1.18–19: "Silva rigens ... | Discolor Usie vultus, sibi dissona massa", with the gloss (f. 183r): "Nota. Hec est differentia inter ypostasim, usiosim, et usiam: ypostasis est materia sine forma, usiosis forma sine materia, usia materia cum forma. Hic vero potentia, usia pro ypostasi abusive." *Cf.* Evrard of Béthune, *Graecismus* X.157–60, ed. Wrobel, 81: "De forma, de materia, de composito que | Dicitur aequiuoce substantia, Graecus at ille | Sic mihi distinguit: substans sit hypostasis ipsa, | Usiosis forma, sed compositum sit usia." The

Brian Stock summarily dismissed the Laudian gloss as "a highly unintelligent commentary on the *Cosmographia*".[93] This is hardly fair. As I have suggested, there are various points at which one might disagree with its readings, but this is natural in a work of such determined difficulty. Overall though the gloss represents an ambitious—and apparently isolated—attempt to get to grips with the varied challenges of interpretation posed by the *Cosmographia*. And as a purely historical document the Laudian gloss has much to tell us about how a classicising twelfth-century author like Bernard was read by his immediate successors.

Quotation and Imitation

When medieval readers opened a copy of the *Cosmographia* they would have had before them a text surrounded by explanatory notes of one kind or another. But we should also remember that these readers would not always have been looking at a complete text. In fifteen of the twenty-three manuscripts I have examined, the work is essentially complete—leaving aside some variations in the catalogue poem, and the fact that only four copies include II.8, Noys's poetic address to her daughters.[94] However, five of the manuscripts include only metrical excerpts from the work. In his discussion of a Berlin manuscript

point of Bernard's line, however, is not that Silva is *usia*, but that she wears the mottled appearance of that higher state.

93 Stock, *Myth and Science*, 11, n. 1. From this Stock infers that the glossator's remark about the Eugenius recital is not to be trusted. But even if Stock's analysis of the intelligence of the gloss were correct, this would not necessarily discredit a piece of historical information that it reports. More pertinent, surely, is the close familiarity with the twelfth-century French schools suggested by the presence in the Waltham booklist of such masters as Adam of Petit-Pont, Robert of Melun, Peter Helias, Ralph of Beauvais, and William of Conches.

94 It has been suggested that the frequent omission of II.8 is due to the fact that it "became corrupted at an early stage" (*Cosmographia*, trans. Wetherbee, 63). But this is far from clear. The most corrupt of the five manuscripts that transmit the poem (BL Sloane 2477, f. 15rv) contains only four obvious errors: two grammatical mistakes (l. 23: "Invenietque" for "Inveniatque", and l. 28: "veniat" for "uniat"); one error of prosody (l. 13: "que Lune" for "Luneque"); and one instance of dittography (l. 35: "sepulta", copied from the line above in place of "paterna"). Medieval scribes were willing to copy poetry far more corrupt than this, such as the storm scene in *Apollonius of Tyre* (see G. A. A. Kortekaas, *The Story of Apollonius, King of Tyre: A Study of its Greek Origin and an Edition of the Two Oldest Latin Recensions* (Leiden: Brill, 2004), 126–29). It seems unlikely that the comparatively light corruption of II.8 would have forced the wholesale omission of a fifty-line poem.

that also contains just four of the nine metres, Constant Mews has proposed that this might be "an earlier version of the text, before the prose sections were added".[95] Yet the Berlin manuscript truncates the *Cosmographia* to a degree of incomprehensibility that seems unlikely to represent an authorial version. We know furthermore from Bernard's dedicatory epistle to Thierry of Chartres that he was concerned to receive Thierry's approval before publication.[96] Given that all six of these manuscripts are anthologies of grammar and verse, it is more likely that the metres they contain were intended as exemplary extracts for students of Latin versification.

Verses from the *Cosmographia* appear in a number of medieval authors, who used them to add authority and elegance to their own compositions. The most highly prized metre was the catalogue poem, and especially its pithy descriptions of animalkind. In a late-thirteenth-century English manuscript, these sections receive the following preface: "Physiologus and Isidore described the natures of living things very fully. In verse, but more briefly, Bernard the Frenchman describes them as follows."[97] As suggested by the reference to the *Physiologus*, and by the copying of these passages alongside a bestiary, pseudo-Hugh of Saint-Victor's *De bestiis et aliis rebus*, these animal verses could be used for moralising purposes. Alexander Neckam, who studied in Paris between 1175 and 1182, incorporated four couplets from Bernard's catalogue—on the swan, the ape, the beaver and the dog—into moralising discussions of the animal world in his *De naturis rerum* (*c.* 1200–1204).[98] These couplets, together with four others from the catalogue poem, reappear in Alexander's supplement to his *Laus divinae sapientiae*, the *Suppletio defectuum* (1216), a poem that

95 C.J. Mews, "Hugh Metel, Heloise and Peter Abelard: The Letters of an Augustinian Canon and the Challenge of Innovation in Twelfth-Century Lorraine", *Viator* 32 (2001): 59–91 (61, n. 7). The manuscript, Berlin, Staatsbibliothek zu Berlin-Preussischer Kulturbesitz, Phillips 1694 [*s.* xii *ex.*], from the abbey of Saint-Arnould de Metz, contains just I.1, I.3, II.10, and II.14.

96 *Cos.*, 96.9–12: "Consilium tamen fuit, ut perfecta minus pagina nomen sui tacuisset auctoris, adeo usque vestro suscepisset iuditio vel egrediendi sententiam vel latendi."

97 Cambridge, University Library Kk 4 25, f. 55rv: "Animantium vero naturas descripserunt plenius Fisiologus et Ysidorus. Porro Bernardus Francus metrice, sed breviter sic depingit."

98 Alexander Neckam, *De naturis rerum libri duo: with the Poem of the Same Author, De laudibus divinae sapientiae*, ed. T. Wright (London: Longman, 1863): *De naturis rerum* I.49 (swan), p. 100 = *Cos.* I.3.449–50; II.129 (ape), p. 211 = *Cos.* I.3.227–28; II.140 (beaver), p. 220 = *Cos.* I.3.229–30; II.157 (dog), p. 252 = *Cos.* I.3.223–24. For the date of *De naturis rerum*, see Hunt, *Schools and the Cloister*, 30.

is saturated with echoes of the *Cosmographia*.[99] Such passages from the catalogue poem were also used for more satirical purposes. Thus in the *c.* 1187 *Verbum adbreviatum*, a treatise on moral theology by the Parisian master Peter the Chanter, one of the same couplets (on the beaver) is quoted to illustrate excessive sumptuousness of clothing.[100] And luxury is also a target of a thirteenth-century metrical satire on the religious orders, probably written in France, which, in a series of strophes *cum auctoritate*, recycles a line from Bernard's fish catalogue, to satirise the gluttony of the Knights Hospitaller.[101]

In line with such moralising and satirical uses, the *Cosmographia* was also mined by authors of preaching tools, to complement the preaching manuals that were starting to be composed around the turn of the thirteenth century. The tripartite *Speculum maius* by the Dominican friar Vincent of Beauvais (*c.* 1190–1264?), which was intended as a quarry of ancient *sententiae* and

---

99 Alexander Neckam, *Suppletio defectuum; Carmina minora*, ed. P. Hochgürtel, CCCM 221 (Turnhout: Brepols, 2008), xciv, 377–78. In the earlier work (*De laudibus divinae sapientiae* VII.355–57, ed. Wright, 481), Alexander had praised Bernard in the following terms: "Depinxit vario Bernardus schemate flores, | Depinxit flores floribus ille novis."

100 Petrus Cantor, *Verbum adbreviatum* I.83 [= I.85 in *PL* 205], ed. M. Boutry, CCCM 196 (Turnhout: Brepols, 2004), 574.172–75. The codices of Peter's work contain a marginal attribution of these verses—"de sancto Bernardo"—that is rightly erased in one of the manuscripts collated by Monique Boutry. Elsewhere one can find quotations from Bernard misattributed—most improbably—to Saint Augustine (*Epitome of Biblical History*, ed. Townsend, 28: "Yle, beatus Augustinus dicit, Silva rigens, deforme chaos, congressio pugnax, | Discolor Usye uultus, sibi dissona massa" = *Cos.* I.1.18–19), and to Hermes Trismegistus (M. H. Malewicz, "*Libellus de efficatia astrologie*: traité astrologique d'Eudes de Champagne, xiie siècle", *Mediaevalia Philosophica Polonorum* 20 (1974): 3–95 (57): "... Mercurius dicens: 'Est igitur elementans natura celum stelleque signifero pervagantes, eoque elementa commoveant ad ingenitas actiones.' " = *Cos.* I.4.7.8–9). This confusion may have arisen from the reading of the *Cosmographia* alongside the Hermetic *Asclepius*, as in the French manuscript Copenhagen, Kongelige Bibliotek, Fabricius 91 4° [s. xii/xiii] (Klibansky and Regen, *Handschriften*, 81), in which Book I of the *Cosmographia* is found alongside Apuleius's *Asclepius* and *De deo Socratis*. Or else Odo's misattribution may have resulted from the use of Bernard's sentence as a gloss on the relevant passage of the *Asclepius* (XXX.315) in a context such as the northern French *Ascelpius* commentary edited by Paolo Lucentini ("*Glosae super Trismegistum*: un commento medievale all' *Asclepius* ermetico", *AHDLMA* 62 (1995): 189–293), which draws extensively on William of Conches and Alan of Lille, but contains only one obvious borrowing from the *Cosmographia* (p. 235, n. 87).

101 R. Düchting, "Versus de ordinibus", *Mittellateinisches Jahrbuch* 16 (1981): 217–22 (221): "Candida crux sapida vult et querit speciosa: | truta rubens, salmo sapidus, prepinguis alosa [*Cos.* I.3.439]. Bernardus." The other *auctores* quoted are Juvenal, Pamphilus, Theodolus, Prudentius, Walter of Châtillon, Aesop, Ovid, Cato and Horace.

exempla for preachers, relied heavily on the *Florilegium Gallicum*, a poetic anthology probably produced in the schools of Orléans.[102] One of its parts, the *Speculum doctrinale*, which Vincent compiled in the 1230s, includes three substantial verse extracts from the *Cosmographia*—all in the sections on *scientia monastica* rather than that on *philosophia naturalis*—and these appear alongside quotations from a small number of other *moderni*, including Hildebert of Lavardin, Matthew of Vendôme, Walter of Châtillon and Geoffrey of Vinsauf. Bernard does not appear in the *Florilegium Gallicum*, which is an anthology of ancient verse, and all three of Vincent's extracts are conflations of verses from the "Microcosmus" in arrangements that I have not encountered in the manuscripts. One concerns the vice of gluttony,[103] one the harmony between body and soul,[104] and a third the opportunity for *contemplatio caeli* afforded man by his upright posture.[105] The first quotation nicely illustrates the way in which *auctores* such as Bernard were absorbed into the discourse of moral theology in this period. But the second and third quotations suggest one of the adverse effects of the transmission of the *Cosmographia* through excerpts in florilegia. For, as we have seen, Bernard's understanding of the body-soul relationship and the purpose of *contemplatio caeli*—as a practical, astrological enterprise rather than the more abstract, mathematical activity of his Platonic and Stoic sources—are two crucial features of the overall meaning of his work. Quoted out of context, these excerpts provide materials for new rhetorical *inventio*, but at the cost of stripping the passages of their distinctive philosophical meaning.

Turning from quotation to imitation, it is clear that from an early date the rhetoricians held up the *Cosmographia* as a model of good poetic practice.

---

102 Rouse, "Florilegia"; S. Schuler, "*Excerptoris morem gerere*: zur Kompilation und Rezeption klassisch-lateinischer Dichter im 'Speculum historiale' des Vinzenz von Beauvais", *Frühmittelalterliche Studien* 29 (1995): 312–48; S. Wenzel, "The Classics in Late-Medieval Preaching", in *Mediaeval Antiquity*, ed. A. Welkenhuysen, H. Braet and W. Verbeke (Leuven: Leuven University Press, 1995), 127–43. See also M. Paulmier-Foucart and M.-C. Duchesne, *Vincent de Beauvais et le grand miroir du monde* (Turnhout: Brepols, 2004), 87–93, who comment (90): "On peut imaginer que cette constitution de florilèges a été un exercice scolaire pour les frères dominicains sous la conduite du *lector* [i.e. Vincent]."

103 *Speculum doctrinale* IV.156 (Douai, 1624; repr. Graz: Akademische Druck- und Verlagsanstalt, 1965), 390 = *Cos.* II.14.88, 91–96, 150–52, 176–78.

104 *Speculum doctrinale* V.108, 465 = *Cos.* II.14.3–4, 7–14; II.6.9–10; II.10.15–22, 25–26. A similar conflation appears in *Speculum naturale* XXXI.49, 2329 (*De animae infusione in embryone*).

105 *Speculum doctrinale* V.65, 441 = *Cos.* II.10.27–32; II.14.9–14, 41–42, 45–46. A fourth reference to the *Cosmographia*, at *Speculum doctrinale* IV.124 (*De praesumptione*), simply reads (370): "Bernardus Silvester in Microcosmo [*sic*]: 'Pudet docuisse Minervam'" = *Cos.* I.1.66.

In his *Ars versificatoria* Matthew of Vendôme quoted two couplets from his master—the only modern cited in the work, with the exception of Matthew himself—in order to illustrate antithesis and the use of epithets.[106] Furthermore Bernard's catalogue poem as a whole provided an exemplary counterpart to Matthew's account of *descriptio loci*, which was a popular topos in poetry of the later twelfth century: the relevant section of Matthew's *Ars* is, for example, found alongside a rearranged text of Bernard's catalogue poem in BN lat. 15155, an anthology of ancient and medieval Latin verse, copied around 1200.[107] English masters also valued Bernard's work. Gervase of Melkley, for example, who had studied in France under John of Hauville, incorporated into his *Ars poetica* (1208–1216) no fewer than sixty illustrative extracts from the *Cosmographia*, the largest tally from any single source. And Gervase's high opinion of Bernard also emerges in his statement that a careful reading of either the *Cosmographia* or his own master's *Architrenius* (among the moderns) should on its own suffice "to inform the untutored mind".[108]

The study of authoritative extracts would often have been a prelude to imitation. Gerald of Wales, for instance, studied in the Parisian schools in the 1160s, and wrote a poem, *De mundi creatione*, that draws heavily on *Cos.* I.1, I.3, and II.10, metres that appear in a contemporary verse anthology, Reims, Bibliothèque municipale 1275. Here is the opening couplet of Gerald's poem:

A tenebris tristique chao, cum caeca teneret
 Congeries rerum semina cuncta simul[109]

---

106 Antithesis by the use of substantive (*Ars* III.28, in *Opera*, ed. Munari, III: 179) = *Cos.* I.3.41–42: "In stellis Codri paupertas, copia Cresi, | Incestus Paridis Ypolitique pudor"; adjectival epithet (*Ars* III.41, ed. Munari, III: 184) = *Cos.* I.3.439–38 [*sic*]: "Turca rubens, salmo sapidus, prepinguis alosa, | Gardo brevis, longus barbulus, ampla plais". Wollin, "Beiträge", argues that parts of Matthew's *Ars* were in circulation before the traditional date of *c.* 1175.

107 The *Cosmographia* extracts (ff. 45v–49v) follow seamlessly from the previous article, interrupted only by the heading "Bernardi Silvestris loci descriptio".

108 Gervase of Melkley, *Ars poetica*, ed. Gräbener, 3.20–26: "Magister Iohannes de Hanvilla, cuius ubera discipline rudem adhuc mihi lactaverunt infantiam, multas quidem elegantias adinvenit, plures auditoribus suis tradidit. In libello vero suo de peregrino philosopho, quem Architrenium vocat, plurimas observavit. Cuius quidem libelli sola sufficit inspectio studiosa rudem animum informare. Idem de Claudiano, de Frigio Darete, de Bernardo Silvestri." See also Chapter 1, 37–38, above.

109 *De mundi creatione et continentis eiusdem*, ll. 9–10, in Gerald of Wales, *Opera*, ed. J. S. Brewer, 8 vols. (London: Longman, 1861–1891), I: 341–49 (341). For analysis, see Moser, *Cosmos of Desire*, 174–79.

—and here the first hexameters of Bernard's:

> Congeries informis adhuc, cum Silva teneret
> Sub veteri confusa globo primordia rerum

It would be interesting to know whether Bernard of Chartres would have thought Gerald had risen here to exemplary imitation (*ad exprimendam auctorum imaginem*) or was merely sewing on cloth from another source.[110] In any case, we may now turn from Gerald's rather straightforward use of the *Cosmographia* to a much more elaborate—and ambivalent—imitation of the work.

Alan of Lille and Peter Lombard

Alan of Lille (d. 1202) was undoubtedly the writer who responded in the most imaginative way to Bernard's *Cosmographia*. The relationship between the two authors is a complex one. It has been suggested that Alan studied with Bernard at Tours.[111] Yet there is no evidence to support this, and indeed in his only direct reference to Bernard, Alan criticises him on a point of grammar.[112] There can be no doubting, however, the depth of Alan's engagement with Bernard's work in both of his major literary works, the *De planctu Naturae* and the *Anticlaudianus*. Both have rightly been characterised as sequels to the *Cosmographia*, due to thematic and verbal parallels between the works, and to Alan's revisions and developments of aspects of Bernard's philosophy. Alan did to Bernard, in a sense, what Bernard did to his own sources: he imitated the *Cosmographia* in ways that would have been obvious to contemporary readers, but his borrowings serve partly to throw into relief the moments at which Alan makes his own distinctive contributions, by correcting doctrines that he held to be erroneous. This dynamic is most evident in Alan's *De planctu Naturae*, a very early response to Bernard that may well have been written within a decade

---

110 See above, nn. 1–2.

111 *Cos.*, 10.

112 Alan of Lille, *Expositio prosae de angelis*, in *Alain de Lille: Textes inédits*, ed. M.-T. d'Alverny (Paris: Vrin, 1965), 194–217 (212), criticises Bernard's use of the plural form of the Hebrew noun *seraph*, in conjunction with a singular verb, at *Cos.* I.3.15: "Quam secus ardescit seraphin…"

of the *Cosmographia*.[113] Like that work, the *De planctu* alternates passages of prose and verse, but it develops the didactic expectations of the mixed form in a different direction, by reconfiguring some of Bernard's natural scientific ideas as less controversial doctrines of Christian ethics.[114]

Since Edmond Faral's classic 1926 study of the intellectual background to the *Roman de la Rose*, there has been a scholarly consensus that one of the principal differences between the allegories of Bernard and Alan is their presentation of the goddess Natura.[115] For whereas Bernard's Natura is a creative, cosmological deity, Alan's is primarily an enforcer of ethical norms. Like another of its prose and verse sources, Boethius's *De consolatione Philosophiae*, the *De planctu* relates a vision in which the poet is approached by the goddess Natura, who descends from the heavens to present a long complaint, occupying most of the work. Where Bernard's Natura, in the opening poem of the *Cosmographia*, had come to God to complain about the disordered state of matter, Alan's Natura addresses herself to man, complaining that he has become estranged from God by lapsing into immoral behaviour. Bernard's goddess is a force for cosmic formation at the beginning of time, whilst Alan's is a force for moral reformation following the Fall.

The similarities and differences between these two approaches are brought out in the chapter in which Alan's Natura begins to explain the reasons for her visit. Her mission is to recall sinful man to his true nature, and she begins her complaint with an account of man's creation that is clearly modelled on the *Cosmographia*. Natura tells man that she is the one "who of old brought your material body into real existence from the mixed substance of primordial

113 The *De planctu* has traditionally been dated *c*. 1160–1170, but an earlier date of 1150–1155 has been proposed on the basis that Alan's work is used in Peter Riga's *Descriptio cuiusdam nemoris* (1155–1160). See C. Wollin, "Der 'Floridus Aspectus' D des Petrus Riga: Erstausgabe nach der Handschrift Douai 825 (Teil I)", *Mittellateinisches Jahrbuch* 43 (2008): 355–91 (384–89).

114 For Alan's use of Bernard in his *Anticlaudianus*, see M. Lemoine, "Alain de Lille et l'école de Chartres", in *Alain de Lille, le docteur universel: philosophie, théologie et littérature au xiie siècle*, ed. J.-L. Solère, A. Vasiliu and A. Galonnier (Turnhout: Brepols, 2005), 47–58 (54–56).

115 E. Faral, "*Le Roman de la Rose* et la pensée française au xiiie siècle", *Revue des deux mondes*, 7th ser. 35 (1926): 430–57. See also R. de Lage, *Alain de Lille: poète du xiie siècle* (Montréal: Institut d'Études Médiévales, 1951), 73–74; Gregory, *Platonismo medievale*, 144–49; Wetherbee, *Platonism*, 188; Economou, *Goddess Natura*, 84; and Pabst, *Prosimetrum*, I: 477–78.

matter".[116] She reminds the poet of the measures she took to save man from "external harassment", by establishing a "marriage" between body and spirit.[117] Thus far Alan's account adheres closely to Bernard's description of the stages of human creation in *Cosmographia* II.13. And yet, whilst Bernard conceived of this conflict between body and spirit as a metaphysical necessity, arising from "the stains and ineradicable evil of Silva" ("maculas et inextricabile Silve | ... malum", *Cos.* II.12.19–20), for Alan the struggle of spirit against flesh is a divinely sanctioned ordeal, intended to test man's resolve against the temptations of the flesh:

> Munera enim empta laboribus iocundius omnibus clarescunt gratuitis. Maioris enim laudis meretur preconia qui laborando munus recipit quam qui reperit ociando.
>
> For rewards obtained from victories shine more fair than all other gifts. Rewards purchased by toil bring more honour than all gifts given gratis.[118]

Thus, the deficiencies of the human condition, which are presented in the *Cosmographia* as a metaphysical necessity, appear in the *De planctu* as familiar tests of moral character, without which virtue would be meaningless.

This contrast is underlined by the authors' divergent handling of the language of exile and segregation. Alan's Natura upbraids man for being a willing exile from the virtues: "What delusion of mind", she asks, "has driven reason into exile?"[119] Since Natura deputed the stewardship of the sublunar realm to Venus all kinds of vices have sprung up, including homosexuality, gluttony, lust, avarice, pride and envy (*De planctu* VIII–XIV). The purpose of Natura's visit is to summon an army of Virtues to wage war on these Vices, and the work concludes with the figure of Genius, Natura's enforcement officer, pronouncing anathema on any man who indulges in vice: "Let him be set apart", Genius declares, "from the council of natural things."[120] The essential point is that man's exile from nature is self-inflicted by his choice of the unnatural vices, but

116 *De planctu Naturae* VI.22–23, ed. N. M. Häring, in *Studi medievali*, 3rd ser. 19 (1978): 797–879 (825): "que olim tui corporis materiam adulterina primordialis materie essentia in uerum esse produxi"; trans. J. J. Sheridan, in *The Plaint of Nature* (Toronto: PIMS, 1980), 117.

117 *De planctu* VI.28–29, ed. Häring, 825: "ab exteriori importunitate"; l. 31 "coniugis".

118 *De planctu* VI.69–72, p. 827; trans. Sheridan, 120.

119 *De planctu* VI.14–15, p. 825: "Que alienatio mentis ... animum exulare coegit?"

120 *De planctu* XVIII.145, p. 878: "a naturalium rerum uniformi concilio segregetur."

that his sinful state may be undone by embracing the virtues of chastity, temperance, humility and generosity, described towards the end of the work (*De planctu* XVI). In the *Cosmographia*, by contrast, the ideals of human reform are characterised not in ethical terms, but in terms of the tranquillity of the superlunar cosmos, from which the evil of Silva has been fully purged ("de celo sideribusque... potentius expurgata", *Cos.* II.5.21.5–6), and to which the soul of man will eventually return, "no longer an unwelcome guest" ("iam non incognitus hospes", II.10.53). As we have seen, the mixed form of the *De planctu* and the *Cosmographia* was traditionally associated, through Boethius and Martianus Capella, with fables of the soul's recollection of knowledge lost at the Fall.[121] Departing from the discussions of the knowledge-to-be-recollected in these sources (the liberal arts in Martianus, the wisdom of Stoic *contemplatio caeli* in Boethius), Bernard suggested that man might alleviate some of the adverse conditions of the fallen world through the natural sciences of medicine and astrology. Alan's work, by contrast, reconceives as essentially moral the range of attainments by which man might counter his fallen state. As Bridget Balint has acutely observed, Alan ultimately rejects the ideal of a life lived exclusively in accordance with Natura, for it was under her rule that the vices had been allowed to flourish.[122] Alan's was a less naturalistic, more theological vision of man's reform in the image of his creator, through the practice of virtue.

This shift from science to ethics in the *De planctu* manifests itself in a number of ways. Urania and Physis, for example, the sisters of Natura, whose roles in the *Cosmographia* underscored man's scientific capabilities, are dispensed with by Alan, who streamlines their creative activities into the single figure of Natura.[123] With them goes the rich conspectus of scientific learning that they represented, to be replaced by the ethical precepts of Natura. Furthermore, although Alan also deploys the notion of man-as-microcosm, the idea that man is made in the image of the cosmos—to which he must return through study—is replaced in the *De planctu* by an analogy of man with the ideal *civitas* or *res publica*, governed by the virtues of wisdom and magnanimity.[124] Even

121 See above, Chapter 2, 57–61.

122 Balint, *Ordering Chaos*, 75: "If you try to replace divine Lady Philosophy, who guided the prisoner Boethius in the direction of theology if not to its heart, with the worldly Natura and her associates, then confusion, error, and sin will result... The *De planctu* is in this sense an anti-*Consolation*: rather than illustrating the way upward, step by step, it exemplifies the difficulties and the labyrinthine ensnarements in nets of vices and rhetoric that await the person who lacks a theological polestar."

123 Wetherbee, *Platonism*, 188.

124 *De planctu* VI.96, 101, p. 828.

in terms of literary analogies, there is a distinct shift from Bernard's suggestion that his rhetorical description of the cosmos mimics the *ornatus mundi* to Alan's infamous analogies between correct grammar and versification on the one hand, and proper, procreative heterosexuality on the other.[125]

George Economou has rightly remarked that in Alan's *De planctu*, the work of a preacher and theologian, many of the suspiciously heterodox cosmological ideas entertained by Bernard are corrected to conform more closely to Christian orthodoxy.[126] Alan jettisons certain theologically problematic topics such as Bernard's emanationist hierarchy of Tugaton, Noys and Endelechia, to be replaced simply by God and Natura, whose respective dominions are demarcated in an explicit manner very different from the allusive terms of the *Cosmographia*.[127] Silva is also absent from the dramatis personae, and Alan, who would later write a treatise attacking Cathar dualism (the *Contra hereticos*, c. 1185–1200), brings the whole question of the pre-existence of matter firmly back into orthodox territory, by insisting that God made the world from nothing and entirely by his own will. In what is clearly a critique of Bernard's cosmology, Alan has Natura declare:

> [Deus] non exterioris instrumenti laborante suffragio, non materie preiacentis auxilio, non indigentie stimulante flagitio, sed solius arbitrarie uoluntatis imperio, mundialis regie ammirabilem speciem fabricauit.

> [God] constructed the marvellous form of the kingdom of the world by the command of his deciding will alone, not by the operative aid of any external factor, not by the help of any pre-existing matter, not at the insistent urgings of any need.[128]

In addition to such emendations, the *De planctu* also finds space for the treatment of doctrines central to Christian theology, which are either entirely absent from the *Cosmographia*, such as the doctrine of grace, or else are handled in an allusive and potentially heterodox manner, such as the Fall.[129] This more explicit Christianity is further supported by a prologue to the

125 See above, Chapter 1, 41–42; and compare *De planctu* I.20–32, ed. Häring, 806–7, and J. Ziolkowski, *Alan of Lille's Grammar of Sex: The Meaning of Grammar to a Twelfth-Century Intellectual* (Cambridge, MA: Harvard University Press, 1985).

126 Economou, *Goddess Natura*, 78.

127 See the clear demarcation of God and nature at *De planctu* VI.140–58, pp. 829–30.

128 *De planctu* VIII.204–7, ed. Häring, 840; trans. Sheridan, 144–45.

129 Newman, *God and the Goddesses*, 66.

De planctu—surviving in only one fourteenth-century manuscript, but considered by its editor to be Alan's own composition—in which the activities of Natura are characterised by a string of biblical allusions to Isaiah and Lamentations.[130]

It was Alan's de-cosmologised, ethical version of the goddess Natura that caught the imagination of late-twelfth-century poets more than Bernard's creative and scientific deity. The Frenchman John of Hauville's *Architrenius* (1184–1185), for example, a satire on the church, the court and the schools, is much more indebted to Alan's works than it is to the *Cosmographia*.[131] In particular, John's characterisation of Natura as a regulator of morals derives from Alan's depiction of the goddess.[132] Even when the *Architrenius* clearly borrows from the *Cosmographia*, these borrowings are transformed through the lens of Alan's intermediary allegories. Thus, Architrenius's prayer to Natura in Book IX contains an invocatory phrase drawn from the *Cosmographia*: "Pace tua, Natura, queror..."[133] However, where in Bernard this invocation comes from Natura's plea for the reformation of Silva, John inserts it into the poet's lament for the evils of the world, which he implores Natura to reform, and the substance of this is drawn from the poet's complaint against Natura in Alan's *Anticlaudianus*. Likewise, in Book II, John writes of Purity that she rules over "an empty court" ("vacua ... aula", II.33, p. 145). John's *vacua aula* relates to and revises the *nuda aula* over which Silva rules in the *Cosmographia* (I.1.56), where the term is likewise a metaphor for the womb. Whereas Bernard used the image to denote a cosmological doctrine (pre-existent matter), John used it as an emblem of the praiseworthy virginity of Purity. Both these allusions illustrate ways in which Bernard's cosmological verse was redeployed in an ethical discourse alien to the original sense of the *Cosmographia*.

If Alan's opposition to some of the philosophical ideas in the *Cosmographia* took the form of a literary *imitatio* that both emulated and emended, then

---

130 F. Hurdy, "Prologus Alani *De Planctu Nature*", *AHDLMA* 55 (1988): 169–85 (184): "Unde Natura olim diademate insignata regali [Isa. 62:3; *cf. De planctu* II.40], facta est quasi vidua, que fuerat domina gentium [Lam. 1:1]. Prosternitur in plateis nec est aliquis iam inventus qui eius causibus miserandis voluerit condolere [*cf.* Lam. 1:2]." These allusions are noted and discussed by J. Jolivet, "La figure de Natura dans le *De planctu Naturae*", in *Alain de Lille*, ed. Solère *et al.*, 127–44 (128).

131 For an assessment of their relative influences, see the Introduction to Johannes de Hauvilla, *Architrenius*, ed. P. G. Schmidt (Munich: Finck, 1974), 80–86, and 384 for a list of parallels with the *Cosmographia*.

132 Johannes de Hauvilla, *Architrenius*, trans. and ed. W. Wetherbee (Cambridge: Cambridge University Press, 1994), xxvi.

133 *Architrenius* IX.192, ed. Schmidt, 23; *cf. Cos.* I.55: "Pace tua, Noys alma, loquar..."

another, more direct form of opposition may be found in the systematic works of Christian theology that were starting to be produced in the northern French schools in the early twelfth century.[134] From the schools of Laon, in the first two decades of the century, to the schools of Paris, where Peter Lombard produced his *Sentences* (*c.* 1155–1157), there was considerable development in the content and format of these theological works.[135] Lying behind and providing important sources for the *Sentences* were systematic works from the previous two decades that laid the foundations for the Lombard's comprehensive treatment of all the principal areas of theology, from the doctrine of God and creation, through Christology, to ethics and the sacraments of the Church. Marcia Colish has convincingly shown that on the subject of creation, some of these systematic theologians, such as Hugh of Saint-Victor and Odo of Lucca, faced up to what she calls "the Chartrian challenge", namely a heterogeneous range of contemporary cosmological doctrines about the nature of matter, divine emanation and the world soul.[136] In order to provide a comprehensive Christian account of creation, twelfth-century theologians had, like the Church Fathers before them, to engage with, if only to refute, these potentially heretical ideas about the origin of the world.

Peter Lombard's *Sentences* display a no-nonsense attitude towards cosmology. As Colish has written, the Lombard had "no sympathy at all for the Chartrian project", and was "not interested in whether the Genesis account has anything in common with philosophical accounts of creation".[137] He began the first *distinctio* of Book II of the *Sentences* by citing Bede on the first verse of Genesis, as recorded in the *Glossa ordinaria*: "He shows that there is one original principle, not several as some have thought."[138] The Lombard then cites Plato's counter-claim that there are three principles of creation—God, the exemplars and matter—and that matter is uncreated and therefore God is an *artifex* rather than a creator.[139] Later in the *distinctio* he refers to the similar error of Aristotle, who conceived of three principles, namely the material, the

134 See Evans, *Old Arts and New Theology*; Leclercq, "Renewal of Theology"; Chenu, "Masters of the Theological 'Science' "; Southern, *Scholastic Humanism*, I: 102–33.

135 Colish, *Peter Lombard*, I: 42–90.

136 Colish, *Peter Lombard*, I: 309–36.

137 Colish, *Peter Lombard*, I: 336, 341.

138 Peter Lombard, *Sententie in IV libris distinctae* II, d. 1, c. 1.1, 3rd ed., revised by I. C. Brady, 2 vols. (Rome: Collegii S. Bonaventurae ad Claras Aquas, 1971–1981), I: 329: "Unum esse rerum principium ostendit, non plura, ut quidam putaverunt."

139 *Sententie* II, d. 1, c. 1.2, ed. Brady, I: 330: "Plato namque tria initia existimavit, Deum scilicet, et exemplar, et materiam; et ipsa increata, sine principio et Deum quasi artificem, non creatorem"; *cf. Timaeus* 48A; Calcidius, *Commentarius* CCCVII, ed. Waszink, 309–10.

formal and the efficient, and argued for the eternity of the world.[140] Both were wrong, Peter Lombard states, in their arguments for the eternity of matter, for the Book of Genesis and the Catholic Church assert that nothing was made before time, which began with the creation of the heavens and Earth.[141] Thus Plato's conception of God as an *artifex* is misleading, for whilst God does fulfil that role, he is primarily a creator (II, d. 1, c. 2). Following a formulation elaborated by Hugh of Saint-Victor (*De sacramentis* I.5.5), Peter Lombard wrote that what the Greeks called chaos is in fact matter that is created but still *in forma confusionis*, awaiting the ordering that will be applied by God.[142] Another cosmological notion that Peter Lombard dismissed as heretical was the idea that the second and third persons of the Trinity were creative "instruments" employed by a divine *artifex*.[143] Without referring to any particular philosophers, the Lombard dismisses wholesale with such statements the natural scientific approach to creation adopted by the northern French *physici*. These arguments, extracted from across the first seventeen distinctions of Book II of the *Sentences*, read like a point-by-point rebuttal of the cosmological creationism of thinkers such as Thierry of Chartres, William of Conches, and Bernard Silvestris.

Of course there was nothing new about these condemnations of philosophical theses about creation; they had been an integral feature of hexameral commentaries since the patristic repudiation of pagan cosmology.[144] Indeed many of the Lombard's arguments were drawn from biblical glosses that derived ultimately from the writings of Ambrose and Augustine. What was

140 *Sententie* II, d. 1, c. 3.4, ed. Brady, I: 331: "Aristotiles vero duo principia dixit, scilicet materiam et speciem, et tertium 'operatorium' dictum; mundum quoque semper esse et fuisse."

141 *Sententie* II, d. 2, c. 2, ed. Brady, I: 338: "Nec etiam tempus factum est ante spiritualem, scilicet angelicam naturam, et ante corporalem, scilicet materiam illam quatuor elementorum confusam."

142 *Sententie* II, d. 2, c. 5, ed. Brady, I: 340: "Sicut enim corporalium materia confusa et permixta, quae secundum graecum chaos dicta est, in illo exordio conditionis primariae et formam confusionis habuit, et non habuit formam distinctionis et discretionis donec postea formaretur atque distinctas reciperet species."

143 *Sententie* II, d. 13, c. 7.2, ed. Brady, I: 394: "Putaverunt quidam haeretici quod Pater velut auctor et artifex, Filio et Spiritu Sancto in rerum operatione quasi instrumento uteretur, ex praedictis verbis errandi occasionem sumentes. Quod velut blasphemum atque sanae doctrinae adversum abicit pia fides."

144 Y. Congar, "Le thème du Dieu-Créateur et les explications de l'hexameron dans la tradition chrétienne', in *L'homme devant Dieu: mélanges offerts à Henri de Lubac*, 3 vols. (Paris: Aubier, 1964), I: 189–222; R. M. Grant, *Miracle and Natural Law in Graeco-Roman and Early Christian Thought* (Amsterdam: North-Holland, 1952), 135–52.

new was the ambition of works such as the *Sentences* to elucidate and reconcile conflicting opinions about a given topic, and thereby offer a systematic academic genre capable of incorporating refinements in the discussion of Christian doctrine.[145] In this sense, it was the genre of the *Sentences*, as much as their doctrinal content, that stood in opposition to a work such as the *Cosmographia*, with its circumspect engagement with ancient cosmology, and its equivocal statements about the pre-existence of matter, the cyclical nature of time, and the immediacy of God's involvement in creation. Poetry such as Bernard's—and indeed Alan's—was designed to stimulate the reader to reflect on the underlying truths concealed beneath the fictive *integumentum* of the text. Like the authoritative poetry of antiquity, such works were intended to offer what John of Salisbury called "an image of all the arts" (*omnium artium imago*).[146] By contrast, Hugh of Saint-Victor, whose *De sacramentis* was a source for Peter Lombard's *Sentences*, argued that all poems were merely "appendages of the arts", which "occasionally touch in a scattered and confused fashion upon some topics lifted out of the arts", but do not substantially contribute to them.[147] The problem, then, was not just one of competing cosmological doctrines, but one of rival methodologies—one of which held the language of myth to be a powerful lens through which to study creation, the other that such language was confused and inconsistent, and would ultimately impede the clear thinking required to reach truth.

Peter Lombard was not himself interested in natural science, but when in the thirteenth century his *Sentences* became the principal textbook of the theology faculty at the University of Paris, it provided its more philosophically-inclined commentators with a framework for incorporating natural science into the discussion of Christian doctrine. Thus, thirteenth-century masters continued, like Bernard, to be interested in questions about the balance of

---

145 B. Lawn, *The Rise and Decline of the Scholastic 'Quaestio Disputata', with Special Emphasis on its Use in the Teaching of Medicine and Science* (Leiden: Brill, 1993), 6–17; Southern, *Scholastic Humanism*, II: 133–47.

146 *Metalogicon* I.24, ed. Hall and Keats-Rohan, 54.32–33; Godman, "*Opus consummatum*".

147 *Didascalicon de studio legendi* III.4, ed. C. H. Buttimer (Washington, DC: Catholic University Press, 1939), 54.10–19: "Appendentia artium sunt quae tantum ad philosophiam spectant, id est, quae in aliqua extra philosophiam materia versantur. Aliquando tamen quaedam ab artibus discerpta sparsim et confuse attingunt, vel si simplex narratio est, viam ad philosophiam praeparant. Huiusmodi sunt omnia poetarum carmina, ut sunt tragoediae, comoediae, satirae, heroica quoque et lyrica, et iambica, et didascalica quaedam, fabulae quoque et historiae, illorum etiam scripta quos nunc philosophos appellare solemus, qui et brevem materiam longis verborum ambagibus extendere consueverunt, et facilem sensum perplexis sermonibus obscurare."

natural and divine causation in Eden and the Fall, and like him they brought philosophical doctrines to bear on these problems. It was just that now these discussions tended to be framed in commentaries on the *Sentences*, rather than in allegories like the *Cosmographia*.[148] And in the faculty of arts, the *quaestiones* and *distinctiones* employed by the theologians proved to be tools well suited to the assimilation of the newly translated Aristotelian *libri naturales*, whose philosophy was more propositional in character than the mythographic Neoplatonism in which Bernard and his contemporaries were steeped.[149] The older philosophy did not disappear entirely. As can be seen from the *Accessus philosophorum* (c. 1230–1240), the *Timaeus* and Boethius's *De consolatione Philosophiae* continued to be studied at the University of Paris.[150] It is significant, however, that in a guide for students written at Paris around the same time, brief treatments of these texts follow a long discussion of Aristotle's *Ethics* within the section on moral philosophy, whereas twelfth-century commentators on the *Timaeus*, working with a smaller range of

148 J. Ziegler, "Medicine and Immortality in Terrestrial Paradise", in *Religion and Medicine in the Middle Ages*, ed. P. Biller and J. Ziegler (Woodbridge, Suff.: York Medieval Press, 2001), 201–42, documents the incorporation of medical sources such as Constantine the African's *Pantegni* and Avicenna's *Liber canonis* into discussion of Eden in commentaries on the *Sentences* from the 1240s.

149 Lawn, *Rise and Decline*, 18–38; Balint, *Ordering Chaos*, 137–68 ("Fabulous Philosophizing after 1170"), esp. 141–45; and see also Thomas Aquinas on Aristotle, *De caelo et mundo* I.10, 279b (*In Aristotelis libros: De caelo et mundo; De generatione et corruptione; Meteorologicorum: expositio S. Thomae Aquinatis*, ed. R. M. Spiazzi (Turin: Marietti, 1952), 109), where Aquinas's critique of the "fables and riddles" expressed about the eternity of the world by "certain philosophers, and especially Plato", alters his named source, Simplicius, who mentions only Heraclitus, in such a way as to suggest that Aquinas may have had in mind what he took to be the teachings of poetically-inclined Platonists of the previous century: "Dicunt autem quidam quod isti poetae et philosophi, et praecipue Plato, non sic intellexerunt secundum quod sonat secundum superficiem verborum; sed suam sapientiam volebant quibusdam fabulis et aenigmaticis locutionibus occultare; et quod Aristotelis consuetudo fuit in pluribus non obiicere contra intellectum eorum, qui erat sanus, ne aliquis ex tali modo loquendi errorem incurreret, sicut dicit Simplicius in commento." Compare Simplicius, *On Aristotle's "On the Heavens 1.10–12"*, trans. R. J. Hankinson (Ithaca, NY: Cornell University Press, 2006), 5: "And Heraclitus, who also purveys his wisdom through riddles, does not mean what most people suppose..."

150 *Accessus philosophorum .vii. artium liberalium*, ll. 780–910, ed. C. Lafleur, *Quatre Introductions à la philosophie au xiiie siècle: textes critiques et étude historique* (Montréal: Institut d'Études Médiévales, 1988), 229–37.

ancient texts, had stressed the work's all-embracing character as a treatise that concerned natural as well as moral philosophy, among other disciplines.[151]

These were not favourable conditions for the *Cosmographia* to play the role of the unifying *opus consummatum* for which it was designed. Instead the work's use as a source of illustrative extracts for the *ars poetriae* and of moralising *sententiae* for preachers and theologians reflects the increasingly compartmentalised character of academic life in the later twelfth century.

A sense of ambivalence about Bernard's work is conveyed by a short poem in the early-thirteenth-century English collection of verse known as the Hunterian anthology. This poem, probably by Gervase of Melkley, is a scholar's prayer for assistance in understanding the new Aristotle. It begins with the recognition that in all the principal divisions of philosophy—logic, ethics and physics—Aristotle has triumphed: "The rock of Parmenides, Socrates' lyre, Plato's *silva*, have ceded to Aristotle."[152] However, the poet continues, the writings of Aristotle's commentators (Porphyry, Avicenna and Alexander of Aphrodisias) are themselves difficult to penetrate, and so he asks God for the assistance of those who teach logic. Logic will provide the foundation from which the philosopher may proceed to the study of nature. Or, as the poet writes of logic:

> Hec prima sororum
> Ianua, quam si quis vultu cognoscit aperto,
> Non fugit hunc Phisis, non hunc Natura repellit,
> Non Urania latet. (ll. 24–27)

---

151 *Le "Guide de l'étudiant" d'un maître anonyme de la faculté des arts de Paris au xiiie siècle*, ed. C. Lafleur and J. Carrier (Québec: Université Laval, 1992), 69–72, with the editors' note (20): "Appendice plutôt artificiel de l'exposé sur la morale 'aristotélicienne', l'étude du *Timée* et de la *Consolation de la philosophie* … ne représente plus guère qu'une survivance du passé, l'intérêt pour ces ouvrages—remis en circulation lors de la renaissance carolingienne—ayant culminé au XIIe siècle, chez les chartrains, avant de décliner ensuite rapidement à cause du prodigieux élargissement de la base textuelle de la philosophie."

152 *Parmenidis rupes*, ll. 1–2, ed. B. Harbert, *A Thirteenth-Century Anthology of Rhetorical Poems: Glasgow MS Hunterian V. 8. 14* (Toronto: PIMS, 1975): "Parmenidis rupes, Socratis lira, silva Platonis | Cessit Aristotili." For the tradition that Parmenides sat alone for years on a rock in Egypt devising the science of dialectic, see R. Klibansky, "The Rock of Parmenides: Mediaeval Views on the Origin of Dialectic", *Mediaeval and Renaissance Studies* 1 (1941–1943): 171–86.

This is the first way in to the
Sisters;[153] Physis flees not nor does Nature shun
The man who sees this clearly, from him Urania
Lies not hidden.

Urania, Natura and Physis are the three goddesses in the *Cosmographia* whose powers and activities symbolise Bernard's conception of natural science informed by contemporary medical and astrological thought. The poet is in two minds. On the one hand, he seems to hold out hope of a synthesis between the allegorical philosophy of the *Cosmographia* and the science of the new Aristotle. On the other, he concedes that the *silva* of Plato (and by extension Bernard) has been superseded by Aristotle's natural philosophy, whose underpinnings are in logic, not mythography.

No other text captures quite so vividly the problems that the *Cosmographia* posed for serious readers of the work. Bernard's extraordinary literary talents had been devoted to a treatise that aimed to synthesise the traditional, but increasingly unfashionable study of the *auctores* with the new sources in natural philosophy that were flowing into the northern French schools. But as the stock of the *auctores* declined, and the new universities tailored themselves to the changing demands of society, the particular synthetic achievement of the *Cosmographia* began to lose its relevance. With the overall status of his vision in question, Bernard's *auctoritas* was reduced to that of a rhetorician. The mixed form of the *Cosmographia*, which had once been a generic indicator of the work's lofty didactic ambitions, now facilitated its deconstruction and rearrangement in those manuals—for rhetoric and preaching—that become such a distinctive feature of intellectual life from the later twelfth century. In *Parmenidis rupes* the sense of intellectual dislocation is palpable. The poet clings to the notion that the secrets of nature lie cloaked in allegory. But Aristotle's natural philosophy means that this idea is in danger of becoming a mere conceit—and with it might go much of the value of poetry as a philosophical medium.

153 The sisters are presumably the seven liberal arts.

Conclusion

The *Cosmographia* addresses one of the most enduring problems of Western philosophy: the place of mankind in the natural world. Bernard's view was that the activity of the human soul was gravely impaired by its incarceration in a material body. At death the soul will return to the tranquillity of its celestial home; but in the meantime, Bernard proposed, man may recapture a measure of superlunar harmony through mastery of his terrestrial environment. Such mastery will rely on an understanding of nature, and this in turn depends on the soul's recollection, though study, of innate scientific knowledge now obscured by its corporeal condition. Bernard believed such science to be grounded in two disciplines: medicine, to prevent physiological disturbances to the soul; and astrology, to predict and prepare for coming events. These branches of natural science are man's principal means of self-protection following his exile from Eden, Earth's one oasis of celestial calm. But these sciences are at best palliative: they treat the effects, not the causes of sublunar discord. In Bernard's cosmos, then, the most assiduously philosophical of men—such as the protagonist of his *Mathematicus*—may also turn out to be the most tragic.

Though timeless in its theme, the *Cosmographia* is in many ways alien to modern sensibilities. An allegory in verse and prose, it is a treatise that requires much unwrapping (as twelfth-century commentators might have put it) to reveal the riches within. My approach in this book has been to use the intellectual history of the twelfth century as a lens to focus on and resolve some of the work's complexities. Bernard was a master in the schools of Tours, a centre with a tradition of scholarship stretching back to the Carolingians, who had brought back into manuscript circulation many of the ancient authorities that were most important for the *Cosmographia*. In its mixed form, and in many details of content, Bernard's work imitated the odysseys of human education described by Boethius's *De consolatione Philosophiae* and Martianus Capella's *De nuptiis Philologiae et Mercurii*. The wisdom of these ancients was highly prized in the twelfth-century schools, and it was through imitating them that Bernard underlined the relevance of his treatise to the contemporary world. For Bernard's adherence to the generic expectations of these well-known models must have made all the more striking his departures from them. In place of the seven liberal arts that form the educational backbone of the *De nuptiis*, Bernard favoured a conception of natural science, represented by the goddess Natura, that embraced the astrology and medicine of the recent Arabo-Latin translations, symbolised by Natura's sisters, Urania and Physis. And instead of

 | DOI 10.1163/9789004268357_007

the therapeutic contemplation of the heavens advocated by the *De consolatione Philosophiae*, Bernard insisted on the more practical uses of medicine and astrology in alleviating the chaos of the sublunar world.

But Bernard's imitation and adaptation of the ancients did not end there. For the *Cosmographia* also imitated those pagan authorities who were thought to have offered previews of Christian doctrine beneath the surface covering (*integumentum*) of their texts. In the decade of the *Cosmographia*'s composition, Peter Abelard and William of Conches had been censured for suggesting that the Platonic world soul was a model for understanding the creative activity of the Holy Spirit. Bernard stepped back from this controversy, to write a treatise that recreated—rather than unwrapped—the *integumenta* of the pagans. In his treatment of the Trinity, the Fall, and the Incarnation, Bernard did not simply equate prophetic and philosophical testimony, but rather formulated a series of interpretative cruxes, which, in the best tradition of commentary on the polysemous masterpieces of antiquity, would incite his readers to consider various doctrinal solutions, and thereby draw nearer to the truth.

The *Cosmographia* is sparing in its references to and quotations from the ancient authorities on which it depends. Bernard operated, as a good imitator must, on the level of verbal and structural allusion, rather than direct citation. One presumably deliberate consequence of this is that the work recaptures some of the archetypal qualities of the ancient creation myths by which it was inspired. The *Cosmographia* responded, nonetheless, to a very particular set of historical circumstances, and it was a response that quickly lost some of its relevance in the changing intellectual climate of the later twelfth century. It would be too simplistic to see this merely as the result of a shift from the predominantly Platonic cosmology of the early Middle Ages to the Aristotelian natural science of the later period. As I suggested in the final chapter of this book, scholars had already begun to turn away from the doctrines and style of the *Cosmographia* for a variety of pedagogical and theological reasons in the decades following the work's publication, well before the Western assimilation of Aristotle's natural philosophy.

At the outset of this book I mentioned Roger Bacon's intriguing idea that philosophy was something that could be completed—*complere* is the verb he uses. Writing in the mid-thirteenth century, Bacon thought that it had been completed twice already in antiquity, once by the patriarchs and prophets and once by Solomon, and that another completion was imminent, through the work of scholars like him who had built on the foundations of Aristotle and his Arabic commentators. But for Bacon philosophy would not be completed by the kind of 'finished' masterpiece (*opus consummatum*) that twelfth-century masters had found in the *Timaeus* and *Aeneid*, and imitated and updated in works like the *Cosmographia*. This prolific writer believed the completion of philosophy to be achievable through a quantity of scholarship, not by the comprehensive quality of the *opus consummatum*. In the second part of his *c*. 1266 *Opus maius*, he notes that Solomon wrote 4,005 books, and Aristotle 1,000. By Bacon's own day, the great translation movement had produced a range of new sources that multiplied several times over the limited corpus of predominantly late antique authorities available to his twelfth-century predecessors. The *Cosmographia* was the product of a culture starting to exhaust the interpretative possibilities of the same few authorities, whereas in the thirteenth century masters could return to the practice of commentary with a renewed vigour.

Yet for all the intellectual space that separates Bernard Silvestris from Roger Bacon, it is striking that the *Cosmographia* anticipates certain trends in later medieval philosophy in ways not evident in earlier medieval works. Bernard's Aristotelian conception of *physica* as a science independent of the mathematical quadrivium, and his emphases on the practical uses of medicine and astrology are both commitments that would be shared by Bacon over a century later. Roger Bacon was dismissive of the philosophical achievements of the twelfth century (*Opus maius* I.15), and his position finds echoes in those modern accounts that have placed a caesura in the intellectual history of the Middle Ages around the year 1200. Part of the interest of the *Cosmographia* is that it warns against such simplified periodisation. It is a work that looks back to earlier traditions of liberal arts scholarship, but also forward to the natural philosophy of the later Middle Ages. This makes the *Cosmographia* an indispensable witness to a period of momentous change in the intellectual history of Europe.

APPENDIX I

Synopsis of the *Cosmographia*

The *Cosmographia* is divided into two books. The first, "Megacosmus", opens with the poetic speech of Natura, who has come before the divine mind, Noys, to complain about the plight of Silva. Silva represents matter, and Natura protests that she has for a long time suffered the unbridled flux of the elements (I.1). Noys agrees to reshape Silva by imposing the bonds of numerical harmony. The four elements emerge from this. United with form, they produce the world, Mundus, who is then betrothed to the world soul, Endelechia, providing the vital spark to animate form and material (I.2). After this mythological opening, Bernard proceeds to an encyclopaedic survey of the cosmos that has been brought into being. This takes the form of a versified catalogue of creation, from the celestial hierarchy, the fixed stars and the planets, down through the climates, mountains and rivers of the Earth, to an extensive list of terrestrial flora and fauna (I.3). "Megacosmus" then concludes with a return to the mythic narrative of the first two chapters, and an explanation of that myth in more strictly philosophical terms, deploying a range of scientific concepts to analyse the relationship between the created world and its ideal model, and the role of the heavens in the generation of life (I.4).

The second book, "Microcosmus", begins with Noys summarising her creative achievements for Natura's approval (II.1–2), and announcing man's creation as the consummation of her work. Natura is to undertake this mission with the help of her sisters, Urania and Physis. Natura then goes in search of Urania, travelling through the cosmos to find her at the outer reaches of the universe (II.3). Urania greets her sister and outlines the task ahead, focusing on her own contribution, namely the knowledge of the heavens that she will instil in the human soul (II.4). The two then proceed to the abode of Tugaton, the *suprema divinitas* who resides at the edge of the universe. Having prayed for the success of their venture, the sisters descend together through the planets, whose individual Oyarses (the personifications of the planets) Bernard describes in graphically mythical terms, and with reference to the nature of their influence on terrestrial affairs (II.5). The goddesses discuss what they see on their journey (II.6). Urania instructs Natura in the nature of the spirits that inhabit the cosmos (II.7), and she recapitulates what Natura has learned so far about the harmonising of the elements, the courses of the planets, meteorology and the link between body and soul (II.8). Next, the goddesses penetrate through the turbulent atmosphere to arrive on Earth at a paradise grove, where they find their sister Physis, accompanied by her two daughters, Theorica and Practica (II.9).

 | DOI 10.1163/9789004268357_008

A Genealogy of Cosmographia *Characters*

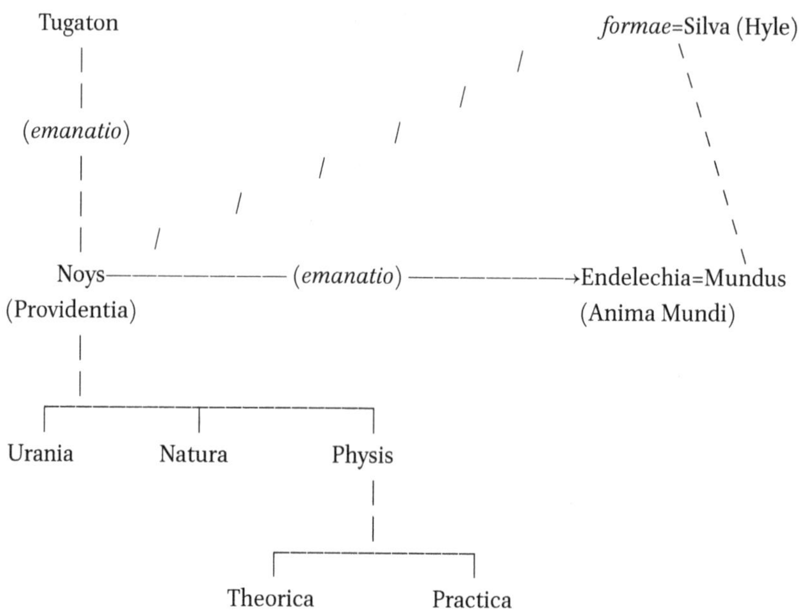

Noys then appears in the grove to deliver a poetic address that evokes the stature and dignity of man (II.10). She assigns specific roles to the three goddesses, and supplies each with an instrument to assist them. Urania is responsible for creating man's soul from Endelechia, and she receives the Mirror of Providence. Physis must create man's body; she is given the Book of Memory. Natura is charged with uniting body and soul, and she will employ the Tablet of Destiny (II.11). The goddesses set about their work, but Physis is perturbed by the difficulty of shaping her material (the turbulent elemental leftovers from the creation of the universe) into a corporeal form (II.12). However, she finally manages to harness the confused matter and, following the example of Noys in the ordering of Silva, resolves elemental disorder into the balanced complexions of human physiology (II.13). Physis then creates man's organs and senses, and the work finishes with the creation of the genitals, whose activity ensures the continuation of mankind (II.14).

APPENDIX II

Census of *Cosmographia* Manuscripts

The following census lists the fifty-three known manuscripts of the *Cosmographia*. In his edition Dronke enumerated a total of forty-six copies of the work, to which he has since added a further four.[1] Two more manuscripts have been noted by Raymond Klibansky[2] and one by Tony Hunt.[3]

The first part of the census tabulates the twenty-three manuscripts used in this study, namely those dating from the twelfth and thirteenth centuries that are currently in French and British libraries. I have extracted from the literature cited under each entry information about the manuscript's date or, where possible, the date of the section that contains the *Cosmographia*, together with the titles of other codicologically related items, and in some cases details of the manuscript's medieval provenance. This bibliography is by no means complete: it includes only the standard manuscript catalogue entry (in some cases over two hundred years old) and any other studies or more recent manuscript descriptions that seemed particularly important. The 'text' column lists the chapters of the *Cosmographia*, including the dedication and prefatory *summa operis*, that each volume contains. It does not report the sometimes significant omissions or interpolations that occur in the long catalogue poem (I.3), much less the minor variants that make each codex unique.

The second part of the census simply lists the remaining thirty *Cosmographia* manuscripts that have not formed part of this study. This census is not intended to be comprehensive, but rather to provide tools for future research on the text and transmission of the *Cosmographia*, and on other topics that may intersect with the work's manuscript tradition.

1 *Cos.*, 64–65; and Dronke, "Bernardo Silvestre": London, Lambeth Palace Library 238; BL Harley 3234; Vatican City, Biblioteca Apostolica Vaticana, Vat. lat. 4357; Seo de Urgel Arch. Cap. 2055.

2 R. Klibansky, "Hidden Treasures at McGill: A Survey of Manuscripts and Historical Documents", *Fontanus* 2 (1989): 65–96 (73): Montreal, McGill University Library, lat. 1; and Klibansky and Regen, *Handschriften*, 81: Copenhagen, Kongelige Bibliotek, Fabricius 91 4°.

3 Hunt, *Teaching and Learning Latin*, I: 157: Cambridge, Gonville and Caius College 593/453.

 | DOI 10.1163/9789004268357_009

I Twelfth and Thirteenth-Century Manuscripts in British and French Libraries

Manuscript	Date	Provenance	Text	Other Contents
Cambridge, Corpus Christi College 406, ff. 65r–74r	xiii *in.*	Oxford University?	Lacks II.8	Seneca, *Tragediae*; John of Hauville, *Architrenius*; Joseph of Exeter, *De bello Troiano*; Geoffrey of Vinsauf, *Poetria nova*; Walter of Châtillon, *Alexandreis*
	M. R. James, *A Descriptive Catalogue of the Manuscripts in the Library of Corpus Christi College, Cambridge*, 2 vols. (Cambridge: Cambridge University Press, 1909–1912), II: 288–91; Rouse, "The A Text of Seneca's *Tragedies*", 114.			
Cambridge, Gonville and Caius College 593/453, ff. 117r–118r	xiii	Cambridge University?	Excerpts from I.3	Grammatical texts, including Priscian's *Institutiones grammaticae* and *De accentibus*, and Theobaldus of Piacenza's *Regula splendescit*
	M. R. James, *A Descriptive Catalogue of the Manuscripts in the Library of Gonville and Caius College*, 2 vols. (Cambridge: Cambridge University Press, 1907–1914), II: 624–27; Hunt, *Teaching and Learning Latin*, I: 157; Hackett, *Original Statutes of Cambridge University*, 348–49.			
Cambridge, Jesus College Q B 17, ff. 118r–132v	xiii	Rievaulx Abbey	Lacks II.8	Cistercian texts, including the *Exordium parvum* and part of the *Carta caritatis prior*
	M. R. James, *A Descriptive Catalogue of the Manuscripts in the Library of Jesus College, Cambridge* (Cambridge: Cambridge University Press, 1895), 43–57; M. L. Dutton, "Ælred of Rievaulx's *Oratio pastoralis*: A New Edition", *Cistercian Studies Quarterly* 38 (2003): 297–308 (297–98).			

Manuscript	Date	Provenance	Text	Other Contents
Cambridge, Trinity College O 7 7, ff. 1r–25r	xiii *in.*	?	Lacks dedication, *summa operis*, and II.8	Anon., *Plato ad ostendendum*; Proba, *Cento Virgilianus de laudibus Christi*; Ovid, *Ibis*; Seneca, *De clementia, De beneficiis, Epistulae*; Jerome, *Adversus Iovinianum* I; Walter Map, *Dissuasio Valerii*

James, *Western Manuscripts in the Library of Trinity College*, III: 348–50; B. Stock, "Hugh of St Victor, Bernard Silvester and MS Trinity College, Cambridge O.7.7", *Mediaeval Studies* 34 (1972): 152–73.

Manuscript	Date	Provenance	Text	Other Contents
Cambridge, Trinity College O 7 40, ff. 71r–119r	xiii	University of Paris?	Complete	Aristotle, *Metaphysica*; Seneca, *Epistulae*; Anon., *Moralium dogma philosophorum*

James, *Western Manuscripts in the Library of Trinity College*, III: 376–79; Ker, *Medieval Libraries of Great Britain*, 47.

Manuscript	Date	Provenance	Text	Other Contents
Cambridge, University Library Kk 4 25, ff. 50rv, 119r–126v	xiii *ex.*	?	Excerpts from I.3; then Book I and first lines of Book II	Pseudo-Hugh of Saint-Victor, *De bestiis et aliis rebus*; Marbod of Rennes, *De lapidibus*; Odo of Meung, *De viribus herbarum*

C. Hardwick and H. R. Luard, *A Catalogue of Manuscripts Preserved in the Library of the University of Cambridge*, 5 vols. (Cambridge: Cambridge University Press, 1856–1867), III: 670–73.

Manuscript	Date	Provenance	Text	Other Contents
London, BL Cotton Cleopatra A XIV, ff. 2r–28r (new foliation)	xiii	?	Lacks *summa operis* and II.8	*Vita Sanctae Margaretae metrica*

TABLE (*Continued*)

Manuscript	Date	Provenance	Text	Other Contents
	A Catalogue of the Manuscripts in the Cottonian Library Deposited in the British Museum (London: Hansard, 1802), 577.			
London, BL Royal 15 A XXXII, ff. 2r–42v	xiii	Merton Priory, Sussex	Lacks II.8	Hildebert of Lavardin, *De querimonia*; Solinus, *De mirabilibus mundi*
	G. F. Warner and J. P. Gilson, *Catalogue of Western Manuscripts in the Old Royal and King's Collections*, 4 vols. (Oxford: Oxford University Press, 1921), II: 152; *Hildeberts Prosimetrum De Querimonia und die Gedichte eines Anonymus: Untersuchungen und kritische Editionen*, ed. P. Orth (Vienna: Österreichische Akademie der Wissenschaften, 2000), 20–21.			
London, BL Sloane 2477, ff. 2r–21r	xiii	?	Complete	Bernard Silvestris, *Mathematicus*; Hildebert of Lavardin, *De querimonia*
	Catalogue of Sloane Manuscripts, 22 vols. (London: British Library, n.d.), VIII: 126–27; *Mathematicus*, ed. Prelog *et al.*, 19; *Mathematicus*, ed. Stone, 217–18; *Hildeberts Prosimetrum*, ed. Orth, 20–21.			
London, Lambeth Palace Library 238, ff. 166r–176r	xiii *in.*	Worcester?	Lacks dedication, II.5.11–22, and II.7–II.9.4	Lawrence of Durham, *Consolatio de morte amici* and other works; Alan of Lille, *Anticlaudianus*; Walter of Châtillon, *Alexandreis*
	M. R. James, *A Descriptive Catalogue of the Manuscripts in the Library of Lambeth Palace: The Medieval Manuscripts* (Cambridge: Cambridge University Press, 1932), 385–91; C. Braun-Irgang, "Tempora nec sexum metuit—ein poetischer Text zum Thema 'De corruptione hominis' ", *Mittellateinisches Jahrbuch* 20 (1985): 128–46.			

<table>
<tr><th>Manuscript</th><th>Date</th><th>Provenance</th><th>Text</th><th>Other Contents</th></tr>
<tr><td>Oxford, Bodl. Digby 157, ff. 1r–15v</td><td>xii</td><td>Battle Abbey, Sussex?</td><td>Lacks II.8</td><td>John of Hauville, Architrenius; Joseph of Exeter, De bello Troiano</td></tr>
<tr><td></td><td colspan="4">R. W. Hunt and A. G. Watson, Bodleian Library Quarto Catalogues, IX: Digby Manuscripts (Oxford: Bodleian Library, 1999), I: 157, II: 71.</td></tr>
<tr><td>Oxford, Bodl. Laud lat. 86, ff. 50v–57v</td><td>xiii</td><td>?</td><td>Excerpts from I.3, II.10, II.14</td><td>Miscellany including the Song of Songs; Anselm, Meditationes; Marbod of Rennes, De lapidibus; Hildebert of Lavardin's poetry; lists of barbarisms and tropes</td></tr>
<tr><td></td><td colspan="4">Coxe, Bodleian Library Quarto Catalogues, II: Laudian Manuscripts (Oxford, 1973), 37–40.</td></tr>
<tr><td>Oxford, Bodl. Laud misc. 515, ff. 182r–219r (new foliation)</td><td>xiii in.</td><td>Waltham Abbey, Essex</td><td>Complete</td><td>Innocent II, De miseria humanae conditionis; Matthew of Vendôme, Tobias; Hildebert of Lavardin, Vita beatae Mariae Egiptiacae; Anon., Vita Sanctae Afrae et Hilariae metrica, Vita Sanctae Katharinae metrica, Sancti Hugonis episcopi Lincolnensis vita; Geoffrey of Vinsauf, Poetria nova</td></tr>
<tr><td></td><td colspan="4">Coxe, Bodleian Library Quarto Catalogues, II: Laudian Manuscripts, 372–74; Ker, Medieval Libraries of Great Britain, 192–93; Mathei Vindocinensis Opera, ed. Munari, I: 80–81; Hunt, Teaching and Learning Latin, I: 51–52.</td></tr>
</table>

TABLE (*Continued*)

Manuscript	Date	Provenance	Text	Other Contents
Oxford, Bodl. Rawlinson G 109, pp. 125–42	xii/xiii	France/England	I.1 and I.3	Around 200 Latin poems, including many by Hugh Primas and Hildebert of Lavardin; with Anon., *De gemellis*; Simon Chèvre d'Or, *Ilias*; Ovid, *Amores* and *Ex Ponto*; and Glanvill, *Tractatus de legibus et consuetudinibus regni Angliae*
	F. Madan *et al.*, *Summary Catalogue of Western Manuscripts in the Bodleian Library*, 7 vols. (Munich: Kraus–Thomson), 432–33 (no. 15479); Wilmart, "Le florilège", 171–72; A. G. Rigg, "Medieval Latin Poetic Anthologies (IV)", *Mediaeval Studies* 43 (1981): 472–97.			
Paris, BN lat. 2770, ff. 146r–148r	xiii	Saint-Martial de Limoges	I.1.1–I.2.13.3; lacks dedication and *summa operis*	Brief theological notes on the sacrament of penance
	P. Lauer *et al.*, *Bibliothèque Nationale: catalogue général des manuscrits latins* (Paris: Bibliothèque Nationale, 1939–), III: 65–67; C. Samaran and R. Marichal, *Catalogue des manuscrits en écriture latine portant indications de date, de lieu ou de copiste, II: Bibliothèque Nationale, fonds latin (Nos 1 à 8,000)* (Paris: CNRS, 1962), 129.			
Paris, BN lat. 6415, ff. 74v–86v.	xii/xiii	Cistercian	Lacks *summa operis* and II.8	William of Conches, *Dragmaticon philosophiae*; Adelard of Bath, *Quaestiones naturales*; Anon., *Omnibus convenit Platonicis*; Martianus Capella, *De nuptiis* VIII; Macrobius, *Commentarii in somnium Scipionis*; Bernard Silvestris, *Mathematicus*; Anon., *De gemellis, De paupere ingrato, De amore*

Manuscript	Date	Provenance	Text	Other Contents
	Catalogus Codicum Manuscriptorum Bibliothecae Regiae, 4 vols. (Paris: Typographia regia, 1739–1744), IV: 241; C. Burnett, "*Omnibus convenit Platonicis*: An Appendix to Adelard of Bath's *Quaestiones naturales*", in *From Athens to Chartres*, ed. Westra: 259–81; B. Eastwood, "Manuscripts of Macrobius, *Commentarii in somnium Scipionis*, before 1500", *Manuscripta* 38 (1994): 138–55 (121); Bernard Silvestris, *Mathematicus*, ed. Prelog *et al.*, 16–17; *Mathematicus*, ed. Stone, 219; William of Conches, *Dragmaticon philosophiae*, ed. Ronca, xxxv, lii–liii.			
Paris, BN lat. 7994, ff. 114v–146v	xiii	?	Complete	Ovid, *Opera*
	Catalogus codicum, IV: 418; F. T. Coulson and B. Roy, *Incipitarium Ovidianum: A Finding Guide for Texts Related to the Study of Ovid in the Middle Ages and Renaissance* (Turnhout: Brepols, 2000), nos. 20, 108, 117, 161, 305, 394, 404, 417.			
Paris, BN lat. 8808 A, ff. 1r–53v	xiii	?	Lacks II.8	No other contents
	Catalogus codicum, IV: 536.			
Paris, BN lat. 14194, ff. 56r–90v	xii/xiii	Saint-Germain-des-Prés	Complete	Hildebert of Lavardin, *De querimonia* and other poems and letters; Pseudo-Vigilius Thapsensis, *De unitate trinitatis*; a glossary of Graecisms
	L. Delisle, *Inventaire des manuscrits de Saint-Germain-des-Prés conservés à la Bibliothèque Impériale, sous les numéros 11504–14231 du fonds latin* (Paris: Auguste Durand et Pédone-Lauriel, 1868), 131–32; *Hildeberts Prosimetrum*, ed. Orth, 11–12.			

TABLE (*Continued*)

Manuscript	Date	Provenance	Text	Other Contents
Paris, BN lat. 15009, ff. 187r–204v	xii	Saint-Victor de Paris	Lacks II.8	Honorius Augustodunensis, *Imago mundi*; Nennius, *Historia Brittonum*; *Abbreviatio Symeonis*; abbreviations of Henry of Huntingdon and Geoffrey of Monmouth; Petrus Alfonsi, *Dialogi contra Iudaeos*
	Honorius Augustodunensis, *Imago mundi*, ed. Flint, 32; Gasparri, "Un copiste lettré"; Ouy, *Les manuscrits*, II: 260–61.			
Paris, BN lat. 15155, ff. 45v–49v	xiii	Saint-Victor de Paris	Excerpts from I.3	Florilegium of classical and medieval Latin verse, including Maximian, Claudian, Statius, Martial, Prudentius, Ovid, Lucan, Hildebert of Lavardin, Matthew of Vendôme and Walter of Châtillon
	Ouy, *Les manuscrits*, II: 558; *Mathei Vindocinensis Opera*, ed. Munari, I: 92–95.			
Reims, Bibliothèque municipale 1275, ff. 178v–181v	xii/xiii	Notre-Dame de Reims	I.1, I.3, II.10, II.14	Verse and prose florilegium, including Hildebert of Lavardin, Marbod of Rennes, Peter Riga and Hugh of Saint-Victor
	H. Loriquet, *Catalogue général des manuscrits des bibliothèques publiques de France: Départements, XXXIX* (Paris: Plon, 1904), 398–437; McDonough, "Hugh Metel".			

Manuscript	Date	Provenance	Text	Other Contents
Troyes, Bibliothèque municipale 787, ff. 201v–217r	xiii/xiv	?	Lacks *summa operis* and II.8	John of Salisbury, *Entheticus minor* and *Policraticus*; Hildebert of Lavardin, *De querimonia*; Walter Map, *Dissuasio Valerii*
	F. Ravaisson *et al.*, *Catalogue général des bibliothèques publiques des départements*, 7 vols. (Paris: Imprimerie Impériale, 1849–1885), II: 324–25; *Hildeberts Prosimetrum*, ed. Orth, 13–14.			

II Other Manuscripts

Auxerre, Bibliothèque municipale 243
Berlin, Staatsbibliothek zu Berlin-Preussischer Kulturbesitz, Phillips 1694 (*olim* Meerm. 180)
Bern, Burgerbibliothek 710
Bern, Burgerbibliothek A 91
Copenhagen, Kongelige Bibliotek, Fabricius 91 4°
Florence, Biblioteca Medicea-Laurenziana, Plut. II 127
Florence, Biblioteca Medicea-Laurenziana, Plut. XXXIII
Kremsmünster, Stiftsbibliothek 149
Montreal, McGill University Library, lat. 1
Munich, Bayerische Staatsbibliothek, clm 23434
Oxford, Bodl. Canon misc. 110
Padova, Biblioteca San Giovanni in Viridario, Plut. destr. XXII
Paris, BN lat. 3245
Paris, BN lat. 6752 A
Paris, BN lat. 8320
Paris, BN lat. 8751 C
Rome, Biblioteca Angelica 401
Seo de Urgel Arch. Cap. 2055
St. Gallen, Kantonsbibliothek Vadiana 1396
Vatican City, Biblioteca Apostolica Vaticana, Pal. lat. 953
Vatican City, Biblioteca Apostolica Vaticana, Reg. lat. 218
Vatican City, Biblioteca Apostolica Vaticana, Reg. lat. 370

Vatican City, Biblioteca Apostolica Vaticana, Reg. lat. 711 A
Vatican City, Biblioteca Apostolica Vaticana, Reg. lat. 1440
Vatican City, Biblioteca Apostolica Vaticana, Vat. lat. 4357
Vienna, Österreichische Nationalbibliothek 283
Vienna, Österreichische Nationalbibliothek 526

Bibliography

Primary Sources

Abelard, Peter. *Historia calamitatum*, ed. J. Monfrin. Paris: Vrin, 1959.

———. *Opera Theologica, II: Theologia christiana; Theologia scholarium; Capitula haeresum Petri Abaelardi*, ed. E. M. Buytaert, CCCM 12. Turnhout: Brepols, 1969.

———. *Opera Theologica, III: Theologia 'summi boni'; Theologia 'scholarium'*, ed. E. M. Buytaert and C. J. Mews, CCCM 13. Turnhout: Brepols, 1987.

———. *Opera Theologica, V: Expositio in Hexameron*, ed. M. Romig, D. Luscombe and C. Burnett, CCCM 15. Turnhout: Brepols, 2004.

Abu Ma'shar. *Liber introductorii maioris ad scientiam judiciorum astrorum*, ed. R. Lemay, 9 vols. Naples: Istituto Universitario Orientale, 1995–1996.

Accessus ad auctores. Bernard d'Utrecht. Conrad d'Hirsau, Dialogus super auctores, ed. R. B. C. Huygens, 2nd ed. Leiden: Brill, 1970.

Accessus philosophorum .vii. artium liberalium, ed. C. Lafleur, *Quatre Introductions à la philosophie au xiiie siècle: textes critiques et étude historique*. Montréal: Institut d'Études Médiévales, 1988: 177–244.

Adelard of Bath. *Conversations with his Nephew: "On the Same and the Different", "Questions of Natural Science", and "On Birds"*, ed. and trans. C. Burnett, with the collaboration of I. Ronca, P. Mantas España and B. van den Abeele. Cambridge: Cambridge University Press, 1998.

Alan of Lille. *Anticlaudianus*, ed. R. Bossuat. Paris: Vrin, 1955.

———. *De planctu Naturae*, ed. N. M. Häring, *Studi medievali*, 3rd ser. 19 (1978): 797–879.

———. *Expositio prosae de angelis*, ed. M.-T. d'Alverny, *Alain de Lille: Textes inédits*. Paris: Vrin, 1965.

———. *The Plaint of Nature*, trans. J. J. Sheridan. Toronto: PIMS, 1980.

———. *Prologus Alani De Planctu Nature*, ed. F. Hurdy, *AHDLMA* 55 (1988): 169–85.

Alcuin of York. *De dialectica*, *PL* 101: 949–76.

———. *Epistolae*, ed. E. Dümmler, *MGH: Epistolae IV*. Berlin: Weidmann, 1895.

An Epitome of Biblical History: Glosses on Walter of Châtillon's Alexandreis 4.176–274, ed. D. Townsend. Toronto: PIMS, 2008.

Anonymus in Boetii Consolationem Philosophiae Commentarius ex Sangallensis Codice, Liber Primus, ed. G. C. Roti. Doctoral Dissertation: State University of New York at Albany, 1979.

Anselm of Laon. *Anselms von Laon systematische Sentenzen*, ed. F. Bliemetzrieder. Münster: Aschendorff, 1919.

Apuleius. *De philosophia libri*, ed. C. Moreschini. Stuttgart: Teubner, 1991.

Aquinas, Thomas. *In Aristotelis libros: De caelo et mundo; De generatione et corruptione; Meteorologicorum: expositio S. Thomae Aquinatis*, ed. R. M. Spiazzi. Turin: Marietti, 1952.

Augustine. *De civitate Dei*, ed. B. Dombart and A. Kalb, 2 vols. 5th ed. Stuttgart: Teubner, 1981.

———. *De Genesi contra Manichaeos*, ed. D. Weber. Vienna: Österreichische Akademie der Wissenschaften, 1998.

———. *Le Genèse au sens littéral en douze livres*, ed. and trans. P. Agaësse and A. Solignac, 2 vols. Paris: Desclée de Brouwer, 1972.

———. *Retractiones*, ed. A. Mutzenbecher, CCSL 57. Turnhout: Brepols, 1984.

Avitus of Vienne. *The Fall of Man: De spiritualis historiae gestis libri i–iii*, ed. D. J. Nodes. Toronto: PIMS, 1985.

Bacon, Roger. *Opus majus*, ed. J. H. Bridges, 3 vols. 2nd ed. London: Williams and Norgate, 1900.

Baudri of Bourgueil. *Poèmes*, ed. and trans. J.-Y. Tilliette, 2 vols. Paris: Les Belles Lettres, 1998–2002.

Bede, pseudo-. *In Mattaei evangelium expositio*, *PL* 92: 9–132.

Bernard of Chartres. *Glosae super Platonem*, ed. P. E. Dutton. Toronto: PIMS, 1991.

Bernard Silvestris. *Cosmographia*, ed. P. Dronke. Leiden: Brill, 1978.

———. *The "Cosmographia" of Bernardus Silvestris*, trans. W. Wetherbee. New York: Columbia University Press, 1973.

———. *De mundi universitate*, ed. C. S. Barach and J. Wrobel. Innsbruck: Wagner'schen Universitaets-Buchhandlung, 1876.

———. *Mathematicus*, ed. and trans. J. Prelog, M. Heim and M. Kiesslich. St Ottilien: EOS, 1993.

———. *Mathematicus*, ed. and trans. D. M. Stone, *AHDLMA* 63 (1996): 209–83.

(pseudo-?) Bernard Silvestris. *The Commentary on Martianus Capella's "De nuptiis Philologiae et Mercurii" Attributed to Bernardus Silvestris*, ed. H. J. Westra. Toronto: PIMS, 1986.

———. *Commentum quod Dicitur Bernardi Silvestris super Sex Libros Eneidos Virgilii*, ed. J. W. Jones and E. F. Jones. Lincoln: University of Nebraska Press, 1977.

Boethius. *Commentarii in librum Aristotelis Peri hermēneias*, ed. C. Meiser. Leipzig: Teubner, 1877.

———. *De consolatione Philosophiae*, ed. L. Bieler, CCSL 94. Turnhout: Brepols, 1957.

———. *De consolatione Philosophiae; Opuscula theologica*, ed. C. Moreschini. Leipzig: Teubner, 2000.

———. *Philosophiae consolationis libri quinque*, ed. R. Peiper. Leipzig: Teubner, 1871.

———. *In Isagogen Porpyhrii commenta*, ed. S. Brandt. Vienna: Tempsky, 1906.

Calcidius. *Commentarius*, ed. J. H. Waszink, *Timaeus a Calcidio translatus commentarioque instructus*. London: Warburg Institute, 1962.

Capitularia Regum Francorum I, MGH: Leges II, ed. A. Boretius. Hannover: Hahn, 1883.

Cassiodorus. *Institutiones*, ed. R. A. B. Mynors. Oxford: Clarendon Press, 1937.

Cicero. *De finibus bonorum et malorum*, ed. C. Moreschini. Munich: Teubner, 2005.

Clarembald of Arras. *Tractatus super librum Boetii de trinitate*, ed. N. M. Häring, *The Life and Works of Clarembald of Arras, a Twelfth-Century Master of the School of Chartres*. Toronto: PIMS, 1965.

Collectio Salernitana, ed. S. de Renzi, 5 vols. Naples: Filiatre-Sebezio, 1852–1859.

Constantine the African. *Opera*, 2 vols. Basel: Heinrich Petri, 1539.

D'Andeli, Henri. *Les "Dits" d'Henri d'Andeli*, ed. A. Corbellari. Paris: Honoré Champion, 2003.

Dominicus Gundissalinus. *De divisione philosophiae*, ed. L. Baur. Münster: Aschendorff, 1903.

Dracontius. *De laudibus Dei*, ed. F. Vollmer, *Fl. Merobardis Reliquae Blosii Aemilii Dracontii Carmina; Eugenii Toletani Episcopi Carmina et Epistulae*. Berlin: MGH, 1905.

Eriugena, Iohannes Scottus. *Periphyseon IV*, ed. É. Jeauneau, CCCM 164. Turnhout: Brepols, 2000.

Evrard of Béthune. *Graecismus*, ed. J. Wrobel. Breslau: Koebner, 1887.

———. *Laborintus*, ed. E. Faral, *Les arts poétiques du xii*[e] *et du xiii*[e] *siècle: recherches et documents sur la technique littéraire du moyen âge*. Paris: Honoré Champion, 1962.

Flores rhetorici, ed. M. Camargo, "A Twelfth-Century Treatise on 'Dictamen' and Metaphor", *Traditio* 47 (1992): 161–213.

Gallia Christiana XIV, ed. B. Hauréau. Paris: Coignard, 1856.

Gerald of Wales. *Opera*, ed. J. S. Brewer, 8 vols. London: Longman, 1861–1891.

Gervase of Melkley. *Ars poetica*, ed. H.-J. Gräbener. Münster: Aschendorff, 1965.

Glosae Colonienses super Macrobium, ed. I. Caiazzo, *Lectures médiévales de Macrobe: les Glosae Colonienses super Macrobium*. Paris: Vrin, 2002.

Glosae super Trismegistum, ed. P. Lucentini, "*Glosae super Trismegistum*: un commento medievale all'*Asclepius* ermetico", *AHDLMA* 62 (1995): 189–293.

Grammatici Latini, ed. H. Keil, 7 vols. Leipzig: Teubner, 1855–1880.

Hermann of Carinthia. *De essentiis*, ed. and trans. C. Burnett. Leiden: Brill, 1982.

Hermes Trismegistus, pseudo-. *De sex rerum principiis*, ed. P. Lucentini and M. D. Delp, CCCM 142. Turnhout: Brepols, 2006.

———. *Liber Hermetis Mercurii Triplicis de vi rerum principiis*, ed. T. Silverstein, *AHDLMA* 22 (1955): 217–302.

Hildebert of Lavardin. *Carmina minora*, ed. A. B. Scott. Leipzig: Teubner, 1969.

———. *Hildeberts Prosimetrum De Querimonia und die Gedichte eines Anonymus: Untersuchungen und kritische Editionen*, ed. P. Orth. Vienna: Österreichische Akademie der Wissenschaften, 2000.

———. *Vita beate Marie Egiptiace*, ed. N. K. Larsen, CCCM 209. Turnhout: Brepols, 2004.

Hildegard of Bingen. *Cause et cure*, ed. L. Moulinier. Berlin: Akademie, 2003.

Hippocrates. *Liber aphorismorum*, in *Articella*. Venice: Baptista de Tortis, 1487: ff. 9r–45v.

Honorius Augustodunensis. *Clavis physicae*, ed. P. Lucentini. Rome: Storia e letteratura, 1974.

———. *Imago mundi*, ed. V. I. J. Flint, *AHDLMA* 49 (1982): 7–151.

Hugh of Bologna. *Rationes dictandi prosaice*, ed. L. Rockinger, *Briefsteller und Formelbücher des eilften bis vierzehnten Jahrhunderts*, 2 vols. Munich: Franz, 1863–1864: I: 47–94.

Hugh of Saint-Victor. *De sacramentis*, *PL* 176: 173–618.

———. *Didascalicon de studio legendi*, ed. C. H. Buttimer. Washington, DC: Catholic University Press, 1939.

———. *The Didascalicon of Hugh of St Victor*, trans. J. Taylor. New York: Columbia University Press, 1961.

Iohannicius. *Isagoge ad Techne Galieni*, ed. G. Maurach, *Sudhoffs Archiv* 62 (1978): 148–74.

Isaac Iudaeus, *Liber dietarum particularium*, in *Omnia Opera Ysaac*. Lyons: Barthélemy Trot, 1515: ff. 103r–156r.

Isidore of Seville. *Etymologiae*, ed. W. M. Lindsay, 2 vols. Oxford: Clarendon Press, 1911.

Jerome. *Epistulae*, ed. I. Hilberg, 4 vols. Vienna: Österreichische Akademie der Wissenschaften, 1996.

John of Hauville. *Architrenius*, ed. P. G. Schmidt. Munich: Finck, 1974.

———. *Architrenius*, trans. and ed. W. Wetherbee. Cambridge: Cambridge University Press, 1994.

John of Salisbury. *Metalogicon*, ed. J. B. Hall and K. S. B. Keats-Rohan, CCCM 98. Turnhout: Brepols, 1991.

———. *Policraticus*, ed. C. C. J. Webb, 2 vols. Oxford: Clarendon Press, 1909.

———. *Policraticus I–IV*, ed. K. S. B. Keats-Rohan, CCCM 118. Turnhout: Brepols, 1993.

John of Wales. *Communiloquium sive summa collationum*. Venice: Georgius de Arrivabenis, 1496.

Lactantius. *De ave phoenice*, ed. and trans. J. Wright and A. M. Duff, *Minor Latin Poets*. Cambridge, MA: Harvard University Press, 1934.

Langton, Stephen. *Quaestiones*, ed. S. Ebbesen and L. B. Mortensen, "A Partial Edition of Stephen Langton's Summa and Quaestiones with Parallels from Andrew Sunesen's Hexaemeron", *Cahiers de l'Institut du moyen-âge grec et latin* 49 (1985): 25–224.

*Le "Guide de l'étudiant" d'un maître anonyme de la faculté des arts de Paris au xiii*e *siècle*, ed. C. Lafleur and J. Carrier. Québec: Université Laval, 1992.

Liber viginti quattuor philosophorum, ed. F. Hurdy, CCCM 143 A. Turhout: Brepols, 1997.

Macrobius. *Commentarii in somnium Scipionis*, ed. J. Willis. Leipzig: Teubner, 1963.

———. *Commentary on the Dream of Scipio*, trans. W. H. Stahl. New York: Columbia University Press, 1952.

Martianus Capella. *De nuptiis Philologiae et Mercurii*, ed. J. Willis. Leipzig: Teubner, 1983.

Matthew of Vendôme. *Opera*, ed. F. Munari, 3 vols. Rome: Storia e letteratura, 1977–1988.

Medieval Literary Theory and Criticism, c. 1100–c. 1375: The Commentary Tradition, ed. A. J. Minnis and A. B. Scott, with the assistance of D. Wallace. Oxford: Clarendon Press, 1988.

Neckam, Alexander. *De naturis rerum libri duo: with the Poem of the Same Author, De laudibus divinae sapientiae*, ed. T. Wright. London: Longman, 1863.

———. *Sacerdos ad altare*, ed. C. J. McDonough, CCCM 227. Turnhout: Brepols, 2010.

———. *Suppletio defectuum; Carmina minora*, ed. P. Hochgürtel, CCCM 221. Turnhout: Brepols, 2008.

Odo of Champagne. *Libellus de efficatia astrologie*, ed. M. H. Malewicz, *Mediaevalia Philosophica Polonorum* 20 (1974): 3–95.

Odo of Meung. *De viribus herbarum*, ed. L. Choulant. Leipzig: Voss, 1832.

Orderic Vitalis. *The Ecclesiastical History of Orderic Vitalis*, ed. and trans. M. Chibnall, 6 vols. Oxford: Clarendon Press, 1969–1980.

Peter Lombard. *Sententie in IV libris distinctae*, 2 vols. 3rd ed., revised by I. C. Brady, Rome: Collegii S. Bonaventurae ad Claras Aquas, 1971–1981.

Peter of Blois. *Carmina*, ed. C. Wollin, CCCM 128. Turnhout: Brepols, 1998.

———. *Libellus de arte dictandi rhetorice*, ed. Camargo, *Medieval Rhetorics*: 37–87.

———. *Opera*, PL 207.

Petrus Cantor. *Verbum adbreviatum*, ed. M. Boutry, CCCM 196. Turnhout: Brepols, 2004.

Plato. *Timaeus*, ed. J. H. Waszink, *Timaeus a Calcidio translatus commentarioque instructus*. London: Warburg Institute, 1962.

Plato ad ostendendum, ed. B. Stock, "Hugh of St Victor, Bernard Silvester and MS Trinity College, Cambridge O.7.7", *Mediaeval Studies* 34 (1972): 152–73.

Priscian. *Praeexercitamina*, ed. Keil, *Grammatici Latini*, III: 430–40.

Quintilian. *Institutio oratoria*, ed. and trans. H. E. Butler, 4 vols. Cambridge, MA: Harvard University Press, 1966–1969.

——— (pseudo-). *Declamationes XIX maiores Quintiliano falso ascriptae*, ed. L. Håkanson. Stuttgart: Teubner, 1982.

Rabanus Maurus. *De universo*, PL 111: 9–614.

Ralph of Longchamps. *In Anticlaudianum Alani commentum*, ed. J. Sulowski. Wrocław: Zakład Narodowy im. Ossolińskich, 1972.

Raymond of Marseille. *Opera Omnia I*, ed. M.-T. d'Alverny, C. Burnett and E. Poulle. Paris: CNRS, 2009.

Recueil de chroniques de Touraine, ed. A. Salmon. Tours: Ladevèze, 1854.

Servius. *De centum metris*, ed. Keil, *Grammatici Latini*, IV: 456–67.

———. *Servii Grammatici qui feruntur in Vergilii carmina commentarii*, ed. G. Thilo and H. Hagen, 3 vols. Leipzig: Teubner, 1881–1887.

Simplicius. *On Aristotle's "On the Heavens 1.10–12"*, trans. R. J. Hankinson. Ithaca, NY: Cornell University Press, 2006.

Stephen of Rouen. *Draco Normannicus*, ed. R. Howlett, *Chronicles of the Reigns of Stephen, Henry II, and Richard I*, 4 vols. London: Longman, 1884–1889.

Sunesen, Andrew. *Hexaemeron*, ed. S. Ebbesen and L. B. Mortensen, 2 vols. Copenhagen: Gad, 1995–1998.

Thierry of Chartres. *Commentaries on Boethius by Thierry of Chartres and his School*, ed. N. M. Häring. Toronto: PIMS, 1971.

Thirteenth-Century Anthology of Rhetorical Poems: Glasgow MS Hunterian V. 8. 14, ed. B. Harbert. Toronto: PIMS, 1975.

"Trotula": A Medieval Compendium of Women's Medicine, ed. and trans. M. H. Green. Philadelphia: University of Pennsylvania Press, 2001.

Versus de ordinibus, ed. R. Düchting, *Mittellateinisches Jahrbuch* 16 (1981): 217–22.

Victorius, Claudius Marius. *Alethia*, ed. J. Martin, *Commodius; Claudius Marius Victorius*, CCSL 128. Turnhout: Brepols, 1960.

Vincent of Beauvais. *Speculum doctrinale*. Douai, 1624; repr. Graz: Akademische Druck- und Verlagsanstalt, 1965.

Walter of Châtillon. *Alexandreis*, ed. M. L. Colker. Padova: Antenore, 1978.

Waltham Chronicle: An Account of the Discovery of our Holy Cross at Montacute and its Conveyance to Waltham, ed. L. Watkiss and M. Chibnall. Oxford: Clarendon Press, 1994.

William of Conches. *Dragmaticon philosophiae*, ed. I. Ronca (with the Catalan *Summa de philosophia vulgari*, ed. L. Badia and J. Pujol), CCCM 152. Turnhout: Brepols, 1997.

———. *Glosae super Boetium*, ed. L. Nauta, CCCM 158. Turnhout: Brepols, 1999.

———. *Glosae super Platonem*, ed. É. Jeauneau, CCCM 203. Turnhout: Brepols, 2006.

———. *Philosophia*, ed. G. Maurach, with the assistance of H. Telle. Pretoria: University of South Africa Press, 1980.

William of Saint-Thierry, *De erroribus Guilelmi de Conchis*, ed. J. Leclercq, *Recueil d'études sur saint Bernard et ses écrits*, 4 vols. Rome: Storia e letteratura, 1962–1987: IV: 357–70.

Secondary Sources

Asper, M. "Silva Parens: zur allegorischen Technik des Bernardus Silvestris", in *Fabrica: Studien zur antiken Literatur und ihrer Rezeption*, ed. T. Baier and F. Schimann. Stuttgart: Teubner, 1997: 129–47.

Baker, A. "The Quest for Guido", *The Private Library* 2 (1969): 139–76.

Balint, B. K. "Hildebert of Lavardin's *Liber de querimonia* in its Cultural Context". Doctoral Dissertation: Harvard University, 2002.

———. *Ordering Chaos: The Self and the Cosmos in Twelfth-Century Latin Prosimetrum.* Leiden: Brill, 2009.

Barker-Benfield, B. C., ed. *St Augustine's Abbey, Canterbury*, 3 vols. London: British Library, 2008.

Baswell, C. "The Medieval Allegorization of the *Aeneid*: MS Cambridge, Peterhouse 158", *Traditio* 41 (1985): 181–237.

———. *Virgil in Medieval England: Figuring the "Aeneid" from the Twelfth Century to Chaucer*. Cambridge: Cambridge University Press, 1995.

Bautier, R.-H. "Chancellerie et culture au Moyen Âge", in *Cancelleria e cultura*, ed. Gualdo: 1–75.

Beaumont, J. "The Latin Tradition of the *De consolatione Philosophiae*", in *Boethius: His Life, Thought, and Influence*, ed. M. Gibson. Oxford: Blackwell, 1981: 278–305.

Bell, D. N., ed. *The Libraries of the Cistercians, Gilbertines and Premonstratensians.* London: British Library, 1992.

Benson, R. L. and G. Constable, with the assistance of C. D. Lanham, ed. *Renaissance and Renewal in the Twelfth Century*. Oxford: Clarendon Press, 1982.

Bezner, F. *Vela Veritatis: Hermeneutik, Wissen und Sprache in der 'Intellectual History' des 12. Jahrhunderts*. Leiden: Brill, 2005.

Bischoff, B. *Manuscripts and Libraries in the Age of Charlemagne*, trans. and ed. Michael Gorman. Cambridge: Cambridge University Press, 1994.

———. *Mittelalterliche Studien: ausgewählte Aufsätze zur Schriftkunde und Literaturgeschichte*, 3 vols. Stuttgart: Hiersemann, 1966–1981.

Boeft, J. den. *Calcidius on Demons (Commentarius ch. 127–136)*. Leiden: Brill, 1977.

Bond, G. A. *The Loving Subject: Desire, Eloquence, and Power in Romanesque France.* Philadelphia: University of Pennsylvania, 1995.

Bonn, B. and A. M. Bautier. *Index Scriptorum Novus Mediae Latinitatis: Supplementum (1973–2005)*. Geneva: Droz, 2005.

Bouchard, C. B. "Monastic Cartularies: Organizing Eternity", in *Charters, Cartularies, and Archives: The Preservation and Transmission of Documents in the Medieval West*, ed. A. J. Kosto and A. Winroth. Toronto: PIMS, 2002: 22–32.

Boutemy, A. "Recherches sur le *Floridus Aspectus* III", *Latomus* 8 (1949): 283–301.

Brague, R. *The Legend of the Middle Ages: Philosophical Explorations of Medieval Christianity, Judaism, and Islam*, trans. L. G. Cochrane. Chicago: University of Chicago Press, 2009.

Brini Savorelli, M. "Il 'Dictamen' di Bernardo Silvestre", *Rivista critica di storia della filosofia* 20 (1965): 182–230.

Brisson, L. "Le discours comme univers et l'univers comme discours: Platon et ses interprètes néo-platoniciens", in *Le texte et sa représentation*. Paris: École Normale Supérieure, 1987: 121–27.

Brühl, C. *Palatium und Civitas: Studien zur Profantopographie spätantiker Civitates vom 3. bis zum 13. Jahrhundert, I: Gallien*, 2 vols. Cologne: Böhlau, 1975–1990.

Bücker, A. "Christianizing the Arts: From Augustine's *De ordine* to Carolingian Thought", in *Signs of Change: Transformations of Christian Traditions and their Representations in the Arts, 1000–2000*, ed. N. H. Petersen, C. Clüver and N. Bell. Amsterdam: Rodopi, 2004: 173–89.

Burnett, C. "The Contents and Affiliation of the Scientific Manuscripts Written at, or Brought to, Chartres in the Time of John of Salisbury", in *World of John of Salisbury*, ed. Wilks: 127–60.

———. "The Establishment of Medieval Hermeticism", in *The Medieval World*, ed. P. Lineham and J. L. Nelson. London: Routledge, 2001: 113–20.

———. *The Introduction of Arabic Learning into England*. London: British Library, 1997.

———. *Magic and Divination in the Middle Ages: Texts and Techniques in the Islamic and Christian Worlds*. Ashgate: Aldershot, 1996.

———. "*Omnibus convenit Platonicis*: An Appendix to Adelard of Bath's *Quaestiones naturales*", in *From Athens to Chartres*, ed. Westra: 259–81.

———. "Physics before the *Physics*: Early Translations from Arabic of Texts Concerning Nature in MSS British Library, Additional 22719 and Cotton Galba E.IV", *Medioevo* 27 (2002): 53–109.

———. "What is the *Experimentarius* of Bernardus Silvestris? A Preliminary Survey of the Material", *AHDLMA* 44 (1977): 79–125.

——— and D. Jacquart, ed. *Constantine the African and ʿAlī ibn al-ʿAbbās al-Magūsī: The "Pantegni" and Related Texts*. Leiden: Brill, 1994.

Bylebyl, J. J. "The Medical Meaning of *Physica*", *Osiris*, 2nd ser. 6 (1990): 16–41.

Caiazzo, I. "Le discussione sull'*Anima mundi* nel secolo xii", *Studi filosofici* 16 (1993): 27–63.

Camargo, M. *Ars dictaminis, Ars dictandi*. Turnhout: Brepols, 1991.

———. "The *Libellus de arte dictandi rhetorice* Attributed to Peter of Blois", *Speculum* 59 (1984): 16–41.

———. *Medieval Rhetorics of Prose Composition: Five English Artes Dictandi and their Tradition*. Binghamton, NY: Medieval and Renaissance Texts and Studies, 1995.

Cameron, A. "Avienus or Avienius?", *Zeitschrift für Papyrologie und Epigraphik* 108 (1995): 252–62.

Carruthers, M. J. *The Book of Memory: A Study of Memory in Medieval Culture*. Cambridge: Cambridge University Press, 1990.

Catalogue of Sloane Manuscripts, 22 vols. London: British Library, n.d.

Catalogue of the Manuscripts in the Cottonian Library Deposited in the British Museum. London: Hansard, 1802.

Catalogus Codicum Manuscriptorum Bibliothecae Regiae, 4 vols. Paris: Typographia regia, 1739–1744.

Chance, J. *Medieval Mythography: From Roman North Africa to the School of Chartres, A.D. 433–1177*. Gainesville: University of Florida Press, 1994.

Chenu, M.-D. "Découverte de la nature et philosophie de l'homme à l'École de Chartres au xiie siècle", *Cahiers d'histoire mondiale* 2 (1954): 313–25.

———. *Nature, Man and Society in the Twelfth Century: New Theological Perspectives in the Latin West*, ed. and trans. J. Taylor and L. K. Little. Chicago: Chicago University Press, 1968.

Clerval, J. A. *Les écoles de Chartres au moyen-age, du ve au xvie siècle*. Chartres: Selleret, 1895.

Colish, M. L. *Peter Lombard*, 2 vols. Leiden: Brill, 1994.

———. *The Stoic Tradition from Antiquity to the Early Middle Ages*, 2 vols. Leiden: Brill, 1985.

Collon, M. *Catalogue général des manuscrits des bibliothèques publiques de France: départements, XXXVII*, 2 vols. Paris: Plon-Nourrit, 1900–1905.

Congar, Y. "Le thème du Dieu-Créateur et les explications de l'hexameron dans la tradition chrétienne", in *L'homme devant Dieu: mélanges offerts à Henri de Lubac*, 3 vols. Paris: Aubier, 1964: I: 189–222.

Contreni, J. J. *Carolingian Learning, Masters and Manuscripts*. Aldershot: Ashgate, 1992.

———. *The Cathedral School of Laon from 850 to 930: Its Manuscripts and Masters*. Munich: Arbeo-Gesellschaft, 1978.

———. "John Scottus, Martin Hiberniensis, the Liberal Arts, and Teaching", in *Insular Latin Studies: Papers on Latin Texts and Manuscripts of the British Isles, 550–1066*, ed. M. W. Herren. Toronto: PIMS, 1981: 23–44.

Copeland, R. "Gloss and Commentary", in *The Oxford Handbook of Medieval Latin Literature*, ed. R. J. Hexter and D. Townsend. Oxford: Oxford University Press, 2012: 171–91.

Cossar, C. D. M. *The German Translations of the Pseudo-Bernhardine "Epistola de cura rei familiaris"*. Göppingen: Alfred Kümmerle, 1975.

Coulson, F. T. and B. Roy. *Incipitarium Ovidianum: A Finding Guide for Texts Related to the Study of Ovid in the Middle Ages and Renaissance*. Turnhout: Brepols, 2000.

Coulter, J. A. *The Literary Microcosm: Theories of Interpretation of the Later Neoplatonists*. Leiden: Brill, 1976.

Courcelle, P. "Tradition platonicienne et traditions chrétiennes du corps-prison", *Revue des études latines* 43 (1965): 406–43.

Courtenay, W. J. "Schools and Schools of Thought in the Twelfth Century", in *Mind Matters: Studies of Medieval and Early Modern Intellectual History in Honour of Marcia Colish*, ed. C. J. Nederman, N. van Deusen, and E. A. Matter. Turnhout: Brepols, 2010: 13–45.

Coxe, H. O. *Bodleian Library, Quarto Catalogues II: Laudian Manuscripts. Reprinted from the Edition of 1858–1885, with Corrections and Additions, and a Historical Introduction,* ed. R. W. Hunt. Oxford: Bodleian Library, 1973.

Curtius, E. R. *Europäische Literatur und lateinisches Mittelalter*. Bern: Francke, 1948.

———. *European Literature and the Latin Middle Ages*, trans. W. R. Trask. New York: Pantheon, 1952.

D'Alverny, M.-T. "Astrologues et théologiens au xiie siècle", in *Mélanges offerts à M.-D. Chenu, Maître en théologie*. Paris: Vrin, 1967: 31–50.

———. *Études sur le symbolisme de la Sagesse et sur l'iconographie*, ed. C. Burnett. Aldershot: Ashgate, 1993.

———. "La Sagesse et ses sept filles: recherches sur les allégories de la Philosophie et des arts libéraux du ixe au xiie siècle", in *Mélanges dediés à la mémoire de Felix Grat*, 2 vols. Paris: Pecqueur-Grat, 1946–1949: I: 245–78.

———. "Translations and Translators", in *Renaissance and Renewal*, ed. Benson and Constable: 421–62.

Dannenfeldt, K. H. "Hermetica philosophica", CTC, I: 137–64.

Dawson, D. *Allegorical Readers and Cultural Revision in Ancient Alexandria*. Berkeley: University of California Press, 1992.

Delisle, L. *Le cabinet des manuscrits de la Bibliothèque Impériale*, 3 vols. Paris: Imprimerie Impériale, 1868–1881.

———. *Inventaire des manuscrits de Saint-Germain-des-Prés conservés à la Bibliothèque Impériale, sous les numéros 11504–14231 du fonds latin*. Paris: Auguste Durand et Pédone-Lauriel, 1868.

———. "Notice sur les manuscrits disparus de la Bibliothèque de Tours pendant la première moitié du xixe siècle", *Notices et extraits des manuscrits* 31 (1883): 157–356.

Delumeau, J. *History of Paradise: The Garden of Eden in Myth and Tradition*, trans. M. O'Connell. New York: Continuum, 1995.

Depreux, P. "La tradition manuscrite des 'Formules de Tours' et la diffusion des modèles d'actes aux viiie et ixe siècles", *Annales de Bretagne et des Pays de l'Ouest* 111 (2004): 55–71.

De Rentis, D. *Die Zeit der Nachfolge: zur Interdependenz von 'imitatio Christi' und 'imitatio auctorum' im 12.–16. Jahrhundert*. Tübingen: Max Niemeyer, 1996.

De Wulf, M. *Histoire de la philosophie médiévale*, 2nd ed. Louvain: Institut Supérieur de Philosophie, 1905.

Diem, A. "The Emergence of Monastic Schools: The Role of Alcuin", in *Alcuin of York*, ed. Houwen and MacDonald: 27–44.

Dolbeau, F. "Un domaine négligé de la littérature médiolatine: les textes hagiographiques en vers", *Cahiers de civilisation médiévale* 178 (2002): 129–39.

Dominik, W. and J. Hall, ed. *A Companion to Roman Rhetoric*. Oxford: Blackwell, 2007.

Donati, M. T. "Metafisica, fisica e astrologia nel xii secolo: Bernardo Silvestre e l'introduzione *qui celum* dell'*Experimentarius*", *Studi Medievali* 31 (1990): 649–703.

Drew, A. "*De eodem et diverso*", in *Adelard of Bath: An English Scientist and Arabist of the Early Twelfth Century*, ed. C. Burnett. London: Warburg Institute, 1987: 17–24.

Dronke, P. "Bernardo Silvestre", in *Enciclopedia Virgiliana*, ed. F. della Corte, 5 vols. Rome: Istituto della Enciclopedia Italiana, 1984–1990: I: 497–500.

———. "Bernard Silvestris, Natura, and Personification", *Journal of the Warburg and Courtauld Institutes* 43 (1980): 16–31.

———. "Eriugena's Earthly Paradise", in *Begriff und Metapher: Sprachform des Denkes bei Eriugena*, ed. W. Beierwaltes. Heidelberg: Winter, 1990: 213–29.

———. *Fabula: Explorations into the Uses of Myth in Medieval Platonism*. Leiden: Brill, 1974.

———, ed. *A History of Twelfth-Century Western Philosophy*. Cambridge: Cambridge University Press, 1988.

———. "New Approaches to the School of Chartres", *Anuario de estudios medievales* 6 (1969): 117–40.

———. *Poetic Individuality in the Middle Ages: New Departures in Poetry, 1000–1150*, 2nd ed. London: Westfield College, 1986.

———. *Sources of Inspiration: Studies in Literary Transformations, 400–1500*. Rome: Storia e letteratura, 1997.

———. *The Spell of Calcidius: Platonic Concepts and Images in the Medieval West*. Florence: SISMEL, 2008.

———. *Verse with Prose from Petronius to Dante: The Art and Scope of the Mixed Form*. Cambridge, MA: Harvard University Press, 1994.

Du Méril, E. *Poésies populaires latines antérieures au douzième siècle*. Paris: Brockhaus et Avenarius, 1843.

Dutton, M. L. "Ælred of Rievaulx's *Oratio pastoralis*: A New Edition", *Cistercian Studies Quarterly* 38 (2003): 297–308.

Dutton, P. E. "The Materialization of Nature and of Quaternary Man in the Early Twelfth Century", in *Man and Nature in the Middle Ages*, ed. S. J. Ridyard and R. G. Benson. Sewanee, TN: University of the South Press, 1995: 137–56.

———. *The Mystery of the Missing Heresy Trial of William of Conches*. Toronto: PIMS, 2006.

Eastwood, B. "Manuscripts of Macrobius, *Commentarii in somnium Scipionis*, before 1500", *Manuscripta* 38 (1994): 138–55.

Economou, G. D. *The Goddess Natura in Medieval Literature*. Cambridge, MA: Harvard University Press, 1972.

Edwards, R. E. "Poetic Invention and the Medieval *Causae*", *Mediaeval Studies* 55 (1993): 183–217.

Elford, D. "William of Conches", in *History of Twelfth-Century Western Philosophy*, ed. Dronke: 308–27.

Else, G. and H. R. Elam. "Imitation", in *The New Princeton Encyclopedia of Poetry and Poetics*, ed. A. Preminger and T. Brogan. Princeton, NJ: Princeton University Press, 1983: 575–79.

Evans, G. R. *Old Arts and New Theology: The Beginnings of Theology as an Academic Discipline*. Oxford: Clarendon Press, 1980.

———. *Problems of Authority in the Reformation Debates*. Cambridge: Cambridge University Press, 1992.

Evans, M. "The *Ysagoge in Theologiam* and the Commentaries Attributed to Bernard Silvestris", *Journal of the Warburg and Courtauld Institutes* 54 (1991): 1–42.

Faral, E. "*Le Roman de la Rose* et la pensée française au xiiie siècle", *Revue des deux mondes*, 7th ser. 35 (1926): 430–57.

Farmer, S. A. *Communities of Saint Martin: Legend and Ritual in Medieval Tours*. Ithaca, NY: Cornell University Press, 1991.

Ferruolo, S. *The Origins of the University: The Schools of Paris and their Critics, 1100–1215*. Stanford, CA: Stanford University Press, 1985.

Fichtenau, H. *Arenga: Spätantike und Mittelalter im Spiegel von Urkundenformeln*. Graz: Böhlau, 1957.

Finckh, R. *Minor Mundus Homo: Studien zur Mikrokosmos-Idee in der mittelalterlichen Literatur*. Göttingen: Vandenhoeck und Ruprecht, 1999.

Flahiff, G. B. "Ecclesiastical Censorship of Books in the Twelfth Century", *Mediaeval Studies* 4 (1942): 1–22.

Flint, V. I. J. *The Rise of Magic in Early Medieval Europe*. Oxford: Clarendon Press, 1991.

———. "The 'School of Laon': A Reconsideration", *Recherches de théologie ancienne et médiévale* 48 (1976): 89–110.

Fontaine, J. *Isidore de Seville et la culture classique de l'Espagne wisigothique*, 3 vols. 2nd ed. Paris: Études Augustiniennes, 1983.

Friis-Jensen, K. and J. M. W. Willoughby, ed. *Peterborough Abbey*. London: British Library, 2001.

Ganz, D. "Mass Production of Early Medieval Manuscripts: The Carolingian Bibles from Tours", in *The Early Medieval Bible: Its Production, Decoration and Use*, ed. R. Gameson. Cambridge: Cambridge University Press, 1994: 53–62.

Gasnault, P. "Les actes privées de l'abbaye de Saint-Martin de Tours du viie au xiie siècle", *Bibliothèque de l'École des Chartes* 112 (1954): 24–66.

———. "Le scriptorium de Saint-Martin à l'époque carolingienne", *Mémoires de la Société Archéologique de Touraine* 62 (1997): 73–86.

Gasparri, F. "Un copiste lettré de l'abbaye de Saint-Victor de Paris au xiie siècle", *Scriptorium* 30 (1976): 232–37.

Gersh, S. "(Pseudo-?) Bernard Silvestris and the Revival of Neoplatonic Virgilian Exegesis", in *Sophies maietores*, ed. Goulet-Cazé *et al.*: 573–93.

Gibson, M. T. *Lanfranc of Bec*. Oxford: Clarendon Press, 1978.

———. "Letters and Charters Relating to Berengar of Tours", in *Auctoritas und Ratio: Studien zu Berengar von Tours*, ed. P. Ganz, R. B. C. Huygens and F. Niewöhner. Wiesbaden: Harrassowitz, 1990: 5–23.

———, D. R. Shanzer and N. F. Palmer. "Manuscripts of Alan of Lille: 'Anticlaudianus' in the British Isles", *Studi medievali*, 3rd ser. 28 (1987): 905–1001.

Gilson, É. "La cosmogonie de Bernardus Silvestris", *AHDLMA* 3 (1928): 5–24.

———. *Dante et la philosophie*. Paris: Vrin, 1939.

———. *La philosophie au moyen âge: des origines patristiques à la fin du xive siècle*, 2nd ed. Paris: Payot, 1944.

Godman, P. "Ambiguity in the *Mathematicus* of Bernardus Silvestris", *Studi medievali*, 3rd ser. 31 (1990): 583–648.

———. "*Opus consummatum, omnium artium… imago*: From Bernard of Chartres to John of Hauvilla", *Zeitschrift für deutsches Altertum und deutsche Literatur* 124 (1995): 26–71.

———. *The Silent Masters: Latin Literature and its Censors in the High Middle Ages*. Princeton, NJ: Princeton University Press, 2000.

Goulet-Cazé, M.-O., G. Madec and D. O'Brien, ed. *Sophies maietores: Chercheurs de sagesse. Hommage à Jean Pépin*. Paris: Institut d'Études Augustiniennes, 1992.

Grandmaison, L. de. *Cartulaire de l'Archevêché de Tours*, 2 vols. Tours: Péricat, 1892–1894.

———. *Procès-verbal du pillage par les Huguenots des reliques et joyaux de Saint-Martin de Tours en mai et juin 1562*. Tours: Mame, 1863.

Grant, R. M. *Miracle and Natural Law in Graeco-Roman and Early Christian Thought*. Amsterdam: North-Holland, 1952.

Gregory, T. "Abélard et Platon", in *Peter Abelard*, ed. E. M. Buytaert. Louvain: Louvain University Press, 1974.

———. *Anima mundi: la filosofia di Guglielmo di Conches e la scuola di Chartres*. Florence: Sansoni, 1955.

———. "La nouvelle idée de nature et de savior scientifique au xiie siècle", in *The Cultural Context of Medieval Learning*, ed. J. E. Murdoch and E. D. Sylla. Dodrecht: Reidel, 1975: 193–218.

———. *Platonismo medievale: studi e ricerche*. Rome: Istituto storico italiano per il Medio Evo, 1958.

Grimm, R. R. *Paradisus coelestis, Paradisus terrestris: zur Auslegungsgeschichte des Paradieses im Abendland bis um 1200*. Munich: Finck, 1977.

Grondeux, A. *Le "Graecismus" d'Évrard de Béthune à travers ses gloses: entre grammaire positive et grammaire spéculative du xiiie au xve siècle*. Turnhout: Brepols, 2000.

Grotans, A. A. *Reading in Medieval St. Gall*. Cambridge: Cambridge University Press, 2006.

Gualdo, G., ed. *Cancelleria e cultura nel Medio Evo*. Vatican City: Archivo Segreto Vaticano, 1990.

Hackett, M. B. *The Original Statutes of Cambridge University: The Text and its History*. Cambridge: Cambridge University Press, 1970.

Hadot, P. *Le voile d'Isis: essai sur l'histoire de l'idée de Nature*. Paris: Gallimard, 2004.

Hardwick, C. and H. R. Luard. *A Catalogue of Manuscripts Preserved in the Library of the University of Cambridge*, 5 vols. Cambridge: Cambridge University Press, 1856–1867.

Haskins, C. H. "An Italian Master Bernard", in *Essays in History Presented to Reginald Lane Poole*, ed. H. W. C. Davis. Oxford: Clarendon Press, 1927: 211–26.

———. *The Renaissance of the Twelfth Century*. Cambridge, Mass.: Harvard University Press, 1927.

———. *Studies in the History of Mediaeval Science*. Cambridge, Mass.: Harvard University Press, 1924.

Hauréau, B. "Maître Bernard", *Bibliothèque de l'École des Chartes* 54 (1893): 792–94.

Haye, T. *Das lateinische Lehrgedicht im Mittelalter: Analyse einer Gattung*. Leiden: Brill, 1997.

———. *Lateinische Oralität: gelehrte Sprache in der mündlichen Kommunikation des hohen und späten Mittelalters*. Berlin: Walter de Gruyter, 2005.

Hildebrandt, M. M. *The External School in Carolingian Society*. Leiden: Brill, 1992.

Histoire littéraire de la France; ouvrage commencé par des religieux bénédictins de la Congrégation de Saint-Maur, et continué par des membres de l'Institut (Académie des inscriptions et belles-lettres), 32 vols. Paris: Imprimerie Nationale, 1835–1981.

Holtz, L. "Alcuin et la renaissance des arts libéraux' ", in *Karl der Grosse und sein Nachwirken*, 5 vols. ed. P. L. Butzer, M. Kerner and W. Oberschelp. Turnhout: Brepols, 1996–1998: I: 45–60.

Houwen, L. A. J. R. and A. A. MacDonald, ed. *Alcuin of York: Scholar at the Carolingian Court*. Groningen: Egbert Forsten, 1998.

Hunt, R. W. *The History of Grammar in the Middle Ages: Collected Papers*, ed. G. L. Bursill-Hall. Amsterdam: John Benjamins, 1980.

———. "The Introductions to the *Artes* in the Twelfth Century", in *Studia Mediaevalia in Honorem Reverendi Patris Raymundi Josephi Martin*. Bruges: De Tempel, 1948: 85–112.

———. *The Schools and the Cloister: The Life and Writings of Alexander Nequam (1157–1217)*, ed. M. Gibson. Oxford: Clarendon Press, 1984.

——— and A. G. Watson. *Bodleian Library Quarto Catalogues, IX: Digby Manuscripts*. Oxford: Bodleian Library, 1999.

Hunt, T. *Teaching and Learning Latin in Thirteenth-Century England*, 3 vols. Cambridge: Brewer, 1991.

Irvine, M. *The Making of Textual Culture: 'Grammatica' and Literary Theory, 350–1100*. Cambridge: Cambridge University Press, 1994.

Jacquart, D. "De *crasis* à *complexio*: note sur le vocabulaire du tempérament en latin médiéval", in *Mémoires, V: textes médicaux latins antiques*, ed. G. Sabbah. Saint-Étienne: Université de Saint-Étienne, 1984: 71–76.

———. "Les emprunts de Guillaume de Conches aux théories médicales", in *Guillaume de Conches*, ed. Obrist and Caiazzo: 79–110.

———. *Science médicale entre deux renaissances (xii*[e] *s.–xv*[e] *s.)*. Aldershot: Ashgate, 1997.

——— and F. Micheau. *La médecine arabe et l'Occident médiéval*. Paris: Maisonneuve et Larose, 1990.

Jaeger, C. S. *The Envy of Angels: Cathedral Schools and Social Ideals in Medieval Europe, 950–1200*. Philadelphia: University of Pennsylvania Press, 1994.

———. "Pessimism in the Twelfth-Century 'Renaissance'", *Speculum* 78 (2003): 1151–83.

James, M. R. *A Descriptive Catalogue of the Manuscripts in the Library of Corpus Christi College Cambridge*, 2 vols. Cambridge: Cambridge University Press, 1909–1912.

———. *A Descriptive Catalogue of the Manuscripts in the Library of Gonville and Caius College*, 2 vols. Cambridge: Cambridge University Press, 1907–1914.

———. *A Descriptive Catalogue of the Manuscripts in the Library of Lambeth Palace: The Medieval Manuscripts*. Cambridge: Cambridge University Press, 1932.

———. *The Western Manuscripts in the Library of Trinity College, Cambridge: A Descriptive Catalogue*, 4 vols. Cambridge: Cambridge University Press, 1900–1904.

Janson, T. *Prose Rhythm in Medieval Latin: From the Ninth to the Thirteenth Century*. Stockholm: Almqvist & Wiksell, 1975.

Jeauneau, É. *L'âge d'or des écoles de Chartres*. Chartres: Houvet, 1995.

———. "'Nani gigantum humeris insidentes': essai d'interprétation de Bernard de Chartres", *Vivarium* 5 (1967): 79–99.

———. "Note sur l'École de Chartres", *Studi medievali*, 3rd ser. 5 (1964): 821–65.

———. *Translatio studii. The Transmission of Learning: A Gilsonian Theme*. Toronto: PIMS, 1995.

———. "L'usage de la notion d'integumentum à travers les gloses de Guillaume de Conches", *AHDLMA* 24 (1975): 35–100.

Jolivet, J. "The Arabic Inheritance", in *History of Twelfth-Century Western Philosophy*, ed. Dronke: 113–48.

———. "La figure de Natura dans le *De planctu Naturae*", in *Alain de Lille*, ed. Solère *et al.*: 127–44.

———. *Perspectives médiévales et arabes*. Paris: Vrin, 2006.

———. "Poésie et philosophie au xiie siècle', *Perspectives médiévales* 17 (1991): 51–70.

———. "Les principes féminins dans la *Cosmographia* de Bernard Silvestre", in *L'homme et son univers au Moyen Âge*, ed. C. Wenin, 2 vols. Louvain: Institut Supérieur de Philosophie, 1986: I: 296–305.

Jones, J. W., Jr. "The So-Called Silvestris Commentary on the *Aeneid* and Two Other Interpretations", *Speculum* 64 (1989): 835–48.

Jones, L. W. "The Art of Writing at Tours from 1000 to 1200 A.D.", *Speculum* 15 (1940): 286–98.

———. "The Script of Tours in the Tenth Century", *Speculum* 14 (1939): 179–98.

Jong, M. de. "From *Scolastici* to *Scioli*: Alcuin and the Formation of an Intellectual Elite", in *Alcuin of York*, ed. Houwen: 45–57.

Kealey, E. J. *Medieval Medicus: A Social History of Anglo-Norman Medicine*. Baltimore, MD: Johns Hopkins University Press, 1981.

Kelly, D. *The Arts of Poetry and Prose*. Turnhout: Brepols, 1991.

———. *The Conspiracy of Allusion: Description, Rewriting, and Authorship from Macrobius to Medieval Romance*. Leiden: Brill, 1999.

Ker, N. R. *Medieval Libraries of Great Britain: A List of Surviving Books*, 2nd ed. London: Royal Historical Society, 1964.

Kim, Y. K. *Augustine's Changing Interpretation of Genesis 1–3: From "De Genesi contra Manichaeos" to "De Genesi ad litteram"*. Lewiston, NY: Edwin Mellen, 2006.

Kindermann, U. *Satyra. Die Theorie der Satire im Mittellateinischen: Vorstudie zu einer Gattungsgeschichte*. Nürnberg: Hans Karl, 1978.

Klaes, M. "Die 'Summa' des Magister Bernardus: zu Überlieferung und Textgeschichte einer zentralen Ars dictandi des 12. Jahrhunderts", *Frühmittelalterliche Studien* 24 (1990): 198–234.

Klibansky, R. "Hidden Treasures at McGill: A Survey of Manuscripts and Historical Documents", *Fontanus* 2 (1989): 65–96.

———. "The Rock of Parmenides: Mediaeval Views on the Origin of Dialectic", *Mediaeval and Renaissance Studies* 1 (1941–1943): 171–86.

——— and F. Regen. *Die Handschriften der philosophischen Werke des Apuleius: ein Beitrag zur Überlieferungsgeschichte*. Göttingen: Vandenhoeck und Ruprecht, 1993.

Köhn, R. "Schulbildung und Trivium im lateinischen Hochmittelalter und ihr möglicher praktischer Nutzen", in *Schulen und Studium im sozialen Wandel des hohen und späten Mittelalters*, ed. J. Fried. Sigmaringen: Thorbecke, 1986: 203–84.

Kortekaas, G. A. A. *Commentary on the Historia Apollonii Regis Tyri*. Leiden: Brill, 2007.

———. *The Story of Apollonius, King of Tyre: A Study of its Greek Origin and an Edition of the Two Oldest Latin Recensions*. Leiden: Brill, 2004.

Krapinger, G. "Die Bienen des armen Mannes in Antike und Mittelalter", in *Theatron. Rhetorische Kultur in Spätantike und Mittelalter / Rhetorical Culture in Late Antiquity and the Middle Ages*. Berlin: Walter de Gruyter, 2007: 189–201.

Kristeller, P. O. "The School of Salerno: Its Development and its Contribution to the History of Learning", *Bulletin of the History of Medicine* 17 (1945): 138–94.

———, F. E. Cranz, V. Brown, J. Hankins, and R. A. Kaster, ed. *Catalogus Translationum et Commentariorum: Mediaeval and Renaissance Latin Translations and Commentaries*. Washington, DC: Catholic University of America Press, 1960–.

Kurdziałek, M. "Der Mensch als Abbild des Kosmos", in *Der Begriff der Repraesentatio im Mittelalter: Stellvertretung, Symbol, Zeichen*, ed. A. Zimmermann. Berlin: Walter de Gruyter, 1971: 35–75.

Lage, R. de. *Alain de Lille: poète du xiie siècle*. Montréal: Institut d'Études Médiévales, 1951.

Lamberton, R. *Homer the Theologian: Neoplatonist Allegorical Reading and the Growth of the Epic Tradition*. Berkeley: University of California Press, 1986.

Landgraf, A. M. *Introduction à l'histoire de la littérature théologique de la scolastique naissante*, ed. and trans. A.-M. Landry and L.-B. Geiger. Montréal: Institut d'Études Médiévales, 1973.

Langlois, C.-V. "Questions d'histoire littéraire: Maître Bernard", *Bibliothèque de l'École des Chartes* 54 (1893): 225–50.

Lapidge, M. "The Stoic Inheritance", in *History of Twelfth-Century Western Philosophy*, ed. Dronke: 81–112.

———. "Versifying the Bible in the Middle Ages", in *The Text in the Community: Essays on Medieval Works, Manuscripts, Authors, and Readers*, ed. J. Mann and M. Nolan. Notre Dame, IN: University of Notre Dame Press, 2006: 11–40.

Lauer, P. *et al. Bibliothèque Nationale: catalogue général des manuscrits latins*. Paris: Bibliothèque Nationale, 1939–.

Lawn, B. *The Rise and Decline of the Scholastic 'Quaestio Disputata', with Special Emphasis on its Use in the Teaching of Medicine and Science*. Leiden: Brill, 1993.

Leclercq, H. "Tours", *Dictionnaire d'archéologie chrétienne et de liturgie*, 15 vols. Paris: Letouzey et Ané, 1924–1953: XV: 2570–677.

Leclercq, J. "The Renewal of Theology", in *Renaissance and Renewal*, ed. Benson and Constable: 68–87.

Lehmann, P. *Mittelalterliche Bibliothekskataloge Deutschlands und der Schweiz*. Munich: Beck, 1918–1939.

Lemay, H. R. "Science and Theology at Chartres: The Case of the Supracelestial Waters", *British Journal for the History of Science* 10 (1977): 226–36.

Lemay, R. *Abu Ma'shar and Latin Aristotelianism in the Twelfth Century: The Recovery of Aristotle's Natural Philosophy through Arabic Astrology*. Beirut: American University, 1962.

———. "Origin and Success of the *Kitab Thamara* of Abu Ja'far Ahmad ibn Yusuf ibn Ibrahim", *Proceedings of the First International Symposium for the History of Arabic Science*, 2 vols. Aleppo: Aleppo University, 1978: II: 91–107.

Lemoine, M. "Alain de Lille et l'école de Chartres", in *Alain de Lille* ed. Solère *et al.*: 47–58.

Lenzen, W. "Überlieferungsgeschichtliche und Verfasseruntersuchen zur lateinischen Liebesdichtung Frankreichs im Hochmittelalter." Doctoral Dissertation: University of Bonn, 1973.

Leonhardt, J. *Dimensio syllabarum: Studien zur lateinischen Prosodie- und Verslehre von der Spätantike bis zur frühen Renaissance*. Göttingen: Vandenhoeck und Ruprecht, 1989.

Lesne, E. *Histoire de la propriété ecclésiastique en France, IV: les livres, 'scriptoria', et bibliothèques du commencement du viiie à la fin du xie siècle*. Lille: Giard, 1938.

———. *Histoire de la propriété ecclésiastique en France, V: les écoles de la fin du viiie siècle à la fin du xiie*. Lille: Giard, 1940.

Lewis, C. S. *The Discarded Image*. Cambridge: Cambridge University Press, 1964.

Lieberg, G. *Poeta Creator: Studien zu einer Figur der antiken Dichtung*. Amsterdam: Gieben, 1982.

Liebeschütz, H. *Mediaeval Humanism in the Life and Writings of John of Salisbury*. London: Warburg Institute, 1950.

Löfstedt, B. "Notizen zur *Cosmographia* des Bernardus Silvestris", *Bulletin du Cange* 51 (1993): 203–8.

Loriquet, H. *Catalogue général des manuscrits des bibliothèques publiques de France: Départements, XXXIX*. Paris: Plon, 1904.

Lucentini, P. "L'Asclepius ermetico nel secolo xii", in *From Athens to Chartres*, ed. Westra: 397–420.

Lutz, C. E. "Remigius' Ideas on the Classification of the Seven Liberal Arts", *Traditio* 12 (1956): 65–86.

———. "Remigius' Ideas on the Origin of the Seven Liberal Arts", *Medievalia et Humanistica* 10 (1956): 32–49.

Mabille, E. "La Pancarte noire de Saint-Martin de Tours, brulée en 1793 et restituée d'après les textes imprimés et manuscrits", *Mémoires de la Société Archéologique de Touraine* 17 (1865): 319–540.

Madan, F. *et al. Summary Catalogue of Western Manuscripts in the Bodleian Library*, 7 vols. Munich: Kraus–Thomson, 1980.

Marenbon, J. "Humanism, Scholasticism and the School of Chartres", *International Journal of the Classical Tradition* 6 (2000): 569–77.

McCrimmon, M. F. "The Classical Philosophical Sources of the *De mundi universitate* of Bernard Silvestris". Doctoral Dissertation: Yale University, 1952.

McDonough, C. J. "Hugh Metel and the Floridus Aspectus of Peter Riga (Staatsbibliothek zu Berlin Preussischer Kulturbesitz Phillips 1694)", *Mediaeval Studies* 67 (2005): 27–74.

McKeon, R. "Medicine and Philosophy in the Eleventh and Twelfth Centuries: The Problem of the Elements", *The Thomist* 24 (1961): 211–56.

———. "Poetry and Philosophy in the Twelfth Century: The Renaissance of Rhetoric", *Modern Philology* 43 (1946): 217–34.

McKitterick, R. *The Carolingians and the Written Word*. Cambridge: Cambridge University Press, 1989.

———. *The Frankish Church and the Carolingian Reforms, 789–895*. London: Royal Historical Society, 1977.

Meier, C. "Die Rezeption des *Anticlaudianus* Alans von Lille in Textkommentierung und Illustration", in *Text und Bild: Aspekte des Zusammenwirkens in Mittelalter und früher Neuzeit*, ed. C. Meier and U. Ruberg. Wiesbaden: Reichert, 1980: 408–548.

Merlet, R. and J. A. Clerval, *Un manuscrit chartrain du xi*[e] *siècle* (Chartres: Garnier, 1893).

Mews, C. J. "Hugh Metel, Heloise and Peter Abelard: The Letters of an Augustinian Canon and the Challenge of Innovation in Twelfth-Century Lorraine", *Viator* 32 (2001): 59–91.

———. "The Lists of Heresies Imputed to Peter Abelard", *Revue bénédictine* 95 (1988): 73–110.

———. *The Lost Love Letters of Heloise and Abelard: Perceptions of Dialogue in Twelfth-Century France*. New York: St. Martin's, 1999.

———. "Nominalism and Theology before Abaelard: New Light on Roscelin of Compiègne", *Vivarium* 30 (1992): 4–33.

———. "St Anselm and Roscelin: Some New Texts and their Implications, I: The *De incarnatione verbi* and the *Disputatio inter Christianum et gentilem*", AHDLMA 66 (1991): 55–98.

Miguel Franco, R. "El texto de la *Epistola de cura rei familiaris* en el *Compendium Morale* de Geremia da Montagnone", *Rivista de Literatura Romanica* 20 (2008): 81–99.

Minkova, M. "Bernardus Silvestris's *Cosmographia* and its Optimism (with an Appendix of Comments and Suggestions on the Text of the *Cosmographia*)", *The Journal of Medieval Latin* 13 (2003): 127–62.

Minnis, A. J. *Medieval Theory of Authorship: Scholastic Literary Attitudes in the Later Middle Ages*. London: Scolar Press, 1984.

Montfaucon, B. de. *Bibliotheca Bibliothecarum Manuscriptorum Nova*, 2 vols. Paris: Briasson, 1739.

Morris, G. *The First Announcement of the Latin Press*. Langford, Som.: Latin Press, 1936.

Morrison, K. F. *The Mimetic Tradition of Reform in the West*. Princeton, NJ: Princeton University Press, 1982.

Moser, T. C., Jr. *A Cosmos of Desire: The Medieval Latin Erotic Lyric in English Manuscripts*. Ann Arbor: University of Michigan Press, 2004.

Moulinier, L. “Abbesse et agronome: Hildegarde et la botanique de son temps”, in *Hildegard of Bingen: The Context of her Thought and Art*, ed. C. Burnett and P. Dronke. London: Warburg Institute, 1998: 135–56.

Munk Olsen, B. *L'étude des auteurs classiques latins aux xie et xiie siècles*, 3 vols. Paris: CNRS, 1982–1989.

Murray, A. *Reason and Society in the Middle Ages*. Oxford: Clarendon Press, 1978.

Nauta, L. “The ‘Glosa’ as Instrument for the Development of Natural Philosophy: William of Conches's Commentary on Boethius”, in *Boethius in the Middle Ages: Latin and Vernacular Traditions of the “Consolatio Philosophiae”*, ed. M. J. F. M. Hoenen and L. Nauta. Leiden: Brill, 1997: 3–39.

Newman, B. *God and the Goddesses: Vision, Poetry, and Belief in the Middle Ages*. Philadelphia: University of Pennsylvania Press, 2003.

Norden, E. *Die antike Kunstprosa vom VI. Jahrhundert v. Chr. bis in die Zeit der Renaissance*, 2 vols. Leipzig: Teubner, 1898.

Obrist, B. and I. Caiazzo, ed. *Guillaume de Conches: philosophie et science au xiie siècle*. Florence: SISMEL, 2011.

O'Donnell, J. R. “The Meaning of ‘Silva’ in the Commentary on the *Timaeus* of Plato by Chalcidius”, *Mediaeval Studies* 7 (1945): 1–20.

Oexle, O. G. *Forschungen zu monastischen und geistlichen Gemeinschaften im westfränkischen Bereich*. Munich: Fink, 1978.

O'Meara, D. J. *Pythagoras Revived: Mathematics and Philosophy in Late Antiquity*. Oxford: Clarendon Press, 1989.

Orlandi, G. *Scritti di filologia mediolatina*, ed. P. Chiesa, A. M. Fagnoni, R. E. Guglielmetti, and G. P. Maggioni. Florence: SISMEL, 2008.

Otten, W. *From Paradise to Paradigm: A Study of Twelfth-Century Humanism*. Leiden: Brill, 2004.

Ouy, G. *Les manuscrits de l'abbaye de Saint-Victor: catalogue établi sur la base du répertoire de Claude de Grandrue (1514)*, 2 vols. Turnhout: Brepols, 1999.

Pabst, B. *Atomtheorien des lateinischen Mittelalters*. Darmstadt: Wissenschaftliche Buchgesellschaft, 1994.

———. *Prosimetrum: Tradition und Wandel einer Literaturform zwischen Spätantike und Spätmittelalter*, 2 vols. Cologne: Böhlau, 1994.

Palmer, N. F. “Plant Names in the *Cosmographia* of Bernardus Silvestris”, *Scientiarum Historia* 20 (1994): 39–56.

Parent, J.-M. *La doctrine de la création dans l'École de Chartres: étude et textes*. Paris: Vrin, 1938.

Paulmier-Foucart, M. and M.-C. Duchesne. *Vincent de Beauvais et le grand miroir du monde*. Brepols: Turnhout, 2004.

Pépin, J. *De la philosophie ancienne à la théologie patristique*. London: Ashgate, 1986.

———. "Exégèse de *In principio* et théorie des principes dans l'*Examerон* (I.4.12–16)", in *Ambrosius Episcopus*, ed. G. Lazatti, 2 vols. Milan: Vita e penserio, 1976: I: 427–82.

———. *Mythe et allégorie: les origines grecques et les contestations judéo-chrétiennes*, 2nd ed. Paris: Études Augustiniennes, 1976.

Peters, E. "Transgressing the Limits Set by the Fathers: Authority and Impious Exegesis in Medieval Thought", in *Christendom and its Discontents: Exclusion, Persecution and Rebellion, 1000–1500*, ed. S. L. Waugh and P. D. Diel. Cambridge: Cambridge University Press, 1996: 338–62.

Pfeiffer, J. *Contemplatio Caeli: Untersuchungen zum Motiv der Himmelsbetrachtung in lateinischen Texten der Antike und des Mittelalters*. Hildesheim: Weidmann, 2001.

Piaia, G. *Vestigia philosophorum: il Medioevo e la storiografia filosofica*. Rimini: Maggioli, 1983.

Pike, D. L. "Bernard Silvestris' Descent into the Classics: The *Commentum super sex libros Aeneidos*", *International Journal of the Classical Tradition* 4 (1998): 343–63.

Poe, J. P. *Caesurae in the Hexameter Line of Latin Elegiac Verse*. Wiesbaden: Steiner, 1974.

Poole, R. L. *Illustrations of the History of Medieval Thought in the Departments of Theology and Ecclesiastical Politics*. London: Williams and Norgate, 1884.

Postles, D. "The Learning of the Austin Canons: The Case of Oseney Abbey", *Nottingham Medieval Studies* 29 (1985): 32–43.

Rand, E. K. *Studies in the Script of Tours*, 2 vols. Cambridge, MA: Mediaeval Academy of America Press, 1929.

Ratkowitsch, C. "Astrologie und Selbstmord im Mathematicus: zu einem Gedicht aus dem Umkreis des Bernardus Silvestris", *Wiener Studien* 112 (1999): 175–229.

———. *Die Cosmographia des Bernardus Silvestris: eine Theodizee*. Cologne: Böhlau, 1995.

———. *Descriptio Picturae: die literarische Funktion der Beschreibung von Kunstwerken in der lateinischen Großdichtung des 12. Jahrhunderts*. Vienna: Österreichische Akademie der Wissenschaften, 1991.

Ravaisson, F. *et al. Catalogue général des bibliothèques publiques des départements*, 7 vols. Paris: Imprimerie Impériale, 1849–1885.

Relihan, J. C. *Ancient Menippean Satire*. Baltimore, MD: Johns Hopkins University Press, 1993.

Reynolds, L. D., ed. *Texts and Transmission: A Survey of the Latin Classics*. Oxford: Clarendon Press, 1983.

Ricklin, T. "Die lateinische Entdeckung der Quintessenz: die *Philosophia* des Daniel von Morley", in *Metaphysics in the Twelfth Century: On the Relationship among Philosophy, Science and Theology*, ed. M. Lutz-Bachmann, A. Fiodora and A. Niederberger. Turnhout: Brepols, 2004: 85–112.

Riddle, J. M. "Dioscorides", in CTC IV: 1–143.

Rigg, A. G. *A History of Anglo-Latin Literature, 1066–1422*. Cambridge: Cambridge University Press, 1992.

———. "Medieval Latin Poetic Anthologies (IV)", *Mediaeval Studies* 43 (1981): 472–97.

Riginos, A. S. *Platonica: The Anecdotes Concerning the Life and Writings of Plato*. Leiden: Brill, 1976.

Ritchey, S. "Rethinking the Twelfth-Century Discovery of Nature", *Journal of Medieval and Early Modern Studies* 39 (2009): 225–55.

Ronca, I. "The Influence of the *Pantegni* on William of Conches's *Dragmaticon*", in *Constantine the African*, ed. Burnett and Jacquart: 266–85.

Rouse, R. H. "The A Text of Seneca's *Tragedies* in the Thirteenth Century", *Revue de l'histoire des textes* 1 (1971): 93–121.

———. "Florilegia and Latin Classical Authors in Twelfth- and Thirteenth-Century Orléans", *Viator* 10 (1979): 131–60.

——— and M. A. Rouse. "The Book Trade at the University of Paris, ca. 1250–ca. 1350", in *La production du livre universitaire au Moyen Age: exemplar et pecia*, ed. L. J. Bataillon, B. G. Guyot and R. H. Rouse. Paris: CNRS, 1988: 41–114.

Russell, D. A. "*De imitatione*", in *Creative Imitation and Latin Literature*, ed. D. West and T. Woodman. Cambridge: Cambridge University Press, 1979: 1–16.

Samaran, C. and R. Marichal. *Catalogue des manuscrits en écriture latine portant indications de date, de lieu ou de copiste, II: Bibliothèque Nationale, fonds latin (Nos 1 à 8,000)*. Paris: CNRS, 1962.

Scafi, A. "Defining *Mappaemundi*", in *The Hereford World Map: Medieval World Maps and their Context*, ed. P. D. A. Harvey. London: British Library, 2006: 345–54.

———. *Mapping Paradise: A History of Heaven on Earth*. Cambridge: Cambridge University Press, 2006.

Schipperges, H. "Einflüsse arabischer Medizin auf die Mikrokosmosliteratur des 12. Jahrhunderts", in *Antike und Orient im Mittelalter*, ed. P. Wilpert and W. P. Eckert. Berlin: Walter de Gruyter, 1971: 129–53.

Schmidt, P. G. "The Commentator Knows Better than the Author", *The Journal of Medieval Latin* 18 (2008): 117–29.

Schuler, S. "*Excerptoris morem gerere*: zur Kompilation und Rezeption klassisch-lateinischer Dichter im 'Speculum historiale' des Vinzenz von Beauvais", *Frühmittelalterliche Studien* 29 (1995): 312–48.

Scott, W. and A. S. Ferguson. *Hermetica: The Ancient Greek and Latin Writings which Contain Religious or Philosophic Teachings Ascribed to Hermes Trismegistus*, 4 vols. Oxford: Clarendon Press, 1924–1936.

Senekovic, D. "Der Anticlaudianus-Kommentar des Radulphus de Longo Campo: zur Kommentierungspraxis im Hochmittelalter", in *Sinnvermittlung: Studien zur Geschichte von Exegese und Hermeneutik*, ed. P. Michel and H. Weder, 2 vols. Zurich: Pano, 2000: I: 475–96.

Shanzer, D. *A Philosophical and Literary Commentary on Martianus Capella's "De nuptiis Philologiae et Mercurii", Book I*. Berkeley: University of California Press, 1986.

Silverstein, T. "*Elementatum*: Its Appearance among the Twelfth-Century Cosmogonists", *Mediaeval Studies* 16 (1954): 156–62.

———. "The Fabulous Cosmogony of Bernardus Silvestris", *Modern Philology* 46 (1948): 92–116.

———. "Review of *The Commentary on the First Six Books of the Aeneid of Vergil Commonly Attributed to Bernardus Silvestris*", ed. J. W. Jones and E. F. Jones (Lincoln, Nebraska: University of Nebraska Press, 1977)", *Speculum* 54 (1979): 154–57.

Siraisi, N. G. *Medieval and Early Renaissance Medicine: An Introduction to Knowledge and Practice*. Chicago: University of Chicago Press, 1990.

Solère, J. L., A. Vasiliu and A. Galonnier, ed. *Alain de Lille, le docteur universel: philosophie, théologie et littérature au xiie siècle*. Turnhout: Brepols, 2005.

Somfai, A. "The Eleventh-Century Shift in the Reception of Plato's *Timaeus* and Calcidius's *Commentary*", *Journal of the Warburg and Courtauld Institutes* 65 (2002): 1–21.

Southern, R. W. *Medieval Humanism and Other Studies*. Oxford: Blackwell, 1970.

———. "The Necessity for Two Peters of Blois", in *Intellectual Life in the Middle Ages: Essays Presented to Margaret Gibson*, ed. L. Smith and B. Ward. London: Hambledon, 1992: 103–18.

———. "Peter of Blois", *Oxford Dictionary of National Biography*, 60 vols. Oxford: Oxford University Press, 2004: VI: 242–47.

———. *Platonism, Scholastic Method and the School of Chartres*. Reading: University of Reading, 1979.

———. *Scholastic Humanism and the Unification of Europe*, 2 vols. Oxford: Blackwell, 1995–2001.

———. "The Schools of Paris and the School of Chartres", in *Renaissance and Renewal*, ed. Benson and Constable: 113–37.

Speer, A. *Die entdeckte Natur: Untersuchungen zu Begründungsversuchen einer 'scientia naturalis' im 12. Jahrhundert* Leiden: Brill, 1995.

Stock, B. *Myth and Science in the Twelfth Century: A Study of Bernard Silvester*. Princeton, NJ: Princeton University Press, 1972.

Stotz, P. "Dichten als Schulfach—Aspekte mittelalterlicher Schuldichtung", *Mittellateinisches Jahrbuch* 16 (1981): 1–16.

Tarabochia Canavero, A. *Esegesi biblica e cosmologia: note sull'interpretazione patristica e medioevale di Genesi 1,2*. Milan: Vita e penserio, 1981.

Tauste Alcocer, F. *Opus naturae: la influencia de la tradición del "Timeo" en la "Cosmographia" de Bernardo Silvestre*. Barcelona: Promociones y Publicaciones Universitarias, 1995.

Teeuwen, M. *Harmony and the Music of the Spheres: The Ars Musica in Ninth-Century Commentaries on Martianus Capella*. Leiden: Brill, 2002.

———. "The Pursuit of Secular Learning: The Oldest Commentary Tradition on Martianus Capella", *The Journal of Medieval Latin* 18 (2008): 36–51.

———. *The Vocabulary of Intellectual Life in the Middle Ages*. Turnhout: Breplos, 2003.

Tester, J. *A History of Western Astrology*. Woodridge, Suff.: Boydell and Brewer, 1987.

Thomson, R. M. *Books and Learning in Twelfth-Century England: The Ending of 'Alter Orbis'*. Walkern, Herts.: Red Gull, 2006.

———. *William of Malmesbury*, 2nd ed. Woodbridge, Suff.: Boydell, 2003.

Thorndike, L. *A History of Magic and Experimental Science*, 8 vols. New York: Columbia University Press, 1923–1958.

Tolomio, I. " 'Corpus carcer' nell'alto medioevo: metamorfosi di un concetto", in *Anima e corpo nella cultura medievale*, ed. C. Casagrande and S. Vecchio. Florence: SISMEL, 1999: 3–19.

Troncarelli, F. *Boethiana Aetas: modelli grafici e fortuna manoscritta della "Consolatio Philosophiae" tra ix e xii secolo*. Alessandria: Edizioni dell'orso, 1987.

Turcan-Verkerk, A.-M. "Le *Prosimetrum* des *Artes dictaminis* médiévales (xiie–xiiie s.)", *Archivum Latinitatis Medii Aevi* 61 (2003): 111–74.

Van Engen, J. "Letters, Schools, and Written Culture in the Eleventh and Twelfth Centuries", in *Dialektik und Rhetorik im früheren und hohen Mittelalter: Rezeption, Überlieferung und gesellschaftliche Wirkung antiker Gelehrsamkeit vornehmlich im 9. und 12. Jahrhundert*, ed. J. Fried. Munich: Oldenbourg, 1997: 97–132.

Van Winden, J. C. M. *Calcidius on Matter: His Doctrines and Sources. A Chapter in the History of Platonism*. Leiden: Brill, 1959.

Vaucelle, E. R. *La collégiale de Saint-Martin de Tours des origines à l'avènement des Valois (397–1328)*. Tours: Péricat, 1908.

Vernet, A. "Bernardus Silvestris et sa *Cosmographia*", in *École Nationale des Chartes. Positions des thèses, 1937*. Nogent-le-Rotrou: Daupeley-Gouverneur, 1937: 167–74.

———. "Poésies latines des xiie et xiiie siècles (Auxerre 243)", in *Mélanges dédiés à la mémoire de Félix Grat*, 2 vols. Paris: Pecqueur-Grat, 1946–1949: II: 271–75.

Von den Steinen, W. "Les sujets d'inspiration chez les poètes latins du xiie siècle", *Cahiers de civilisation médiévale* 9 (1966): 165–75, 363–83.

Von Moos, P. *Geschichte als Topik: das rhetorische Exemplum von der Antike zur Neuzeit und die historiae im "Policraticus" Johanns von Salisbury*. Hildesheim: Oms, 1998.

———. *Hildebert von Lavardin, 1056–1133: Humanitas an der Schwelle des höfischen Zeitalters*. Stuttgart: Anton Hiersemann, 1965.

———. "The Use of *Exempla* in the *Policraticus* of John of Salisbury", in *World of John of Salisbury*, ed. Wilks: 207–61.

Vulliez, C. "L'apprentissage de la rédaction des documents diplomatiques à travers l'ars dictaminis français (et spécialement Ligérien) du xiie siècle", in *Cancellaria e cultura*, ed. Gualdo: 77–96.

Waddell, H. *Peter Abelard*. London: Constable, 1933.

———. *Sylvestris Discourses*. Langford, Som.: Latin Press, 1936.

———. *The Wandering Scholars*. London: Constable, 1927.

Wahlgren, L. *The Letter Collections of Peter of Blois: Studies in the Manuscript Tradition*. Gothenburg: Acta Universitatis Gothoburgensis, 1993.

Wallach, L. *Alcuin and Charlemagne: Studies in Carolingian History and Literature*. Ithaca, NY: Cornell University Press, 1959.

Ward, J. O. *Ciceronian Rhetoric in Treatise, Scholion and Commentary*. Turnhout: Brepols, 1995.

———. "The Date of the Commentary on Cicero's *De inventione* by Thierry of Chartres (*ca.* 1095–1160?) and the Cornifician Attack on the Liberal Arts", *Viator* 3 (1973): 219–73.

———. "Quintilian and the Rhetorical Revolution of the Middle Ages", *Rhetorica* 13 (1995): 231–84.

———. "Rhetorical Theory and the Rise and Decline of *Dictamen* in the Middle Ages and Early Renaissance", *Rhetorica* 19 (2001): 175–223.

Warner, G. F. and J. P. Gilson. *Catalogue of Western Manuscripts in the Old Royal and King's Collections*, 4 vols. Oxford: Oxford University Press, 1921.

Webber, T. *Scribes and Scholars at Salisbury Cathedral, c. 1075–c. 1125*. Oxford: Clarendon Press, 1992.

——— and A. G. Watson, ed. *The Libraries of the Augustinian Canons*. London: British Library, 1998.

Wedel, T. O. *The Medieval Attitude toward Astrology*. New Haven, CT: Yale University Press, 1920.

Wenzel, S. "The Classics in Late-Medieval Preaching", in *Mediaeval Antiquity*, ed. A. Welkenhuysen, H. Braet and W. Verbeke. Leuven: Leuven University Press, 1995: 127–43.

Westra, H. J., ed. *From Athens to Chartres: Neoplatonism and Medieval Thought. Studies in Honour of Edouard Jeauneau*. Leiden: Brill, 1992.

Wetherbee, W. "Philosophy, Cosmology and the Twelfth-Century Renaissance", in *History of Twelfth-Century Western Philosophy*, ed. Dronke: 21–53.

———. *Platonism and Poetry in the Twelfth Century: The Literary Influence of the School of Chartres*. Princeton, NJ: Princeton University Press, 1972.

———. "The Study of the Classical Authors: From Late Antiquity to the Twelfth Century", in *The Cambridge History of Literary Criticism, II: The Middle Ages*, ed. A. Minnis and I. Johnson. Cambridge: Cambridge University Press, 2005: 99–144.

Whitaker, J. "Goodness, Wisdom, Power: A Middle Platonic Triad", in *Sophies maietores*, ed. Goulet-Cazé *et al.*: 179–94.

Whitman, J. *Allegory: The Dynamics of an Ancient and Medieval Technique*. Oxford: Clarendon Press, 1987.

Wickersheimer, E. *Les manuscrits latins de médecine du haut moyen âge dans les bibliothèques de France*. Paris: CNRS, 1966.

Wieland, G. A. "The Glossed Manuscript: Classbook or Library Book?", *Anglo-Saxon England* 14 (1985): 153–73.

Wilks, M., ed. *The World of John of Salisbury*. Oxford: Blackwell, 1984.

Williams, J. R. "The *Microcosmographia* of Trier MS 1041", *Isis* 22 (1934): 106–35.

Wilmart, A. "Le florilège de Saint-Gatien: contribution à l'étude des poèmes d'Hildebert et de Marbode", *Revue Bénédictine* 48 (1936): 3–40, 147–81, 235–58

Wollin, C. "Beiträge zur Werkchronologie und Rezeption des Matthäus von Vendôme", *Sacris Erudiri* 45 (2006): 327–52.

———. "Das Epitaphium Bernardi Siluestris in der Handschrift Angers B. M. 303", *Sacris Erudiri* 42 (2003): 369–401.

———. "Der 'Floridus Aspectus' D des Petrus Riga: Erstausgabe nach der Handschrift Douai 825 (Teil I)", *Mittellateinisches Jahrbuch* 43 (2008): 355–91.

Woods, M. C. *Classroom Commentaries: Teaching the "Poetria nova" across Medieval and Renaissance Europe*. Columbus: Ohio State University Press, 2010.

Woolsey, R. B. "Bernard Silvester and the Hermetic *Asclepius*", *Traditio* 6 (1948): 340–44.

Worstbrock, F. J. "Die Frühzeit der Ars dictandi in Frankreich", in *Pragmatische Schriftlichkeit im Mittelalter: Erscheinungsformen und Entwicklungsstufen*, ed. H. Keller, K. Grubmüller and N. Staubach. Munich: Fink, 1992: 131–56.

———, M. Klaes and J. Lütten. *Repertorium der Artes dictandi des Mittelalters, I: von den Anfängen bis um 1200*. Munich: Fink, 1992.

Ziegler, J. "Medicine and Immortality in Terrestrial Paradise", in *Religion and Medicine in the Middle Ages*, ed. P. Biller and J. Ziegler. Woodbridge, Suff.: York Medieval Press, 2001: 201–42.

Ziolkowski, J. *Alan of Lille's Grammar of Sex: The Meaning of Grammar to a Twelfth-Century Intellectual*. Cambridge, MA: Harvard University Press, 1985.

———. "Cultures of Authority in the Long Twelfth Century", *Journal of English and Germanic Philology* 108 (2009): 421–48.

———. "The Highest Form of Compliment: *Imitatio* in Medieval Latin Culture", in *Poetry and Philosophy in the Middle Ages: A Festschrift for Peter Dronke*, ed. J. Marenbon. Leiden: Brill, 2001: 293–307.

Index of Manuscripts

General Index

www.ingramcontent.com/pod-product-compliance
Lightning Source LLC
LaVergne TN
LVHW010657300926
849161LV00005B/1150
* 9 7 8 9 0 0 4 2 5 6 9 1 0 *